THE ANALYSIS OF NATURAL WATERS

VOLUME 1

COMPLEX–FORMATION PRECONCENTRATION TECHNIQUES

The Analysis of Natural Waters

VOLUME 1

COMPLEX-FORMATION PRECONCENTRATION TECHNIQUES

T. R. CROMPTON

Oxford New York Tokyo
OXFORD UNIVERSITY PRESS
1993

Oxford University Press, Walton Street, Oxford OX2 6DP
Oxford New York Toronto
Delhi Bombay Calcutta Madras Karachi
Kuala Lumpur Singapore Hong Kong Tokyo
Nairobi Dar es Salaam Cape Town
Melbourne Auckland Madrid
and associated companies in
Berlin Ibadan

Oxford is a trade mark of Oxford University Press

Published in the United States
by Oxford University Press Inc., New York

A catalogue record for this book is available from the British Library

Library of Congress Cataloging in Publication Data
Crompton, T.R. (Thomas Roy)
The analysis of natural waters / T.R. Crompton.
p. cm.
Includes bibliographical references and indexes.
Contents: v. 1. Complex-formation preconcentration techniques—
v. 2. Direct preconcentration techniques.
1. Water—Analysis. 2. Water chemistry. I. Title.
TD380.C75 1993 628.1′61—dc20 92-23860

ISBN 0-19-855394-3 Vol. 2
ISBN 0-19-855395-1 Vol. 1
ISBN 0-19-855752-3 (2 Vol. set).

Typeset by Colset Pte. Ltd., Singapore
Printed in Great Britain by
Bookcraft Ltd.,
Midsomer Norton, Avon

PREFACE

This ~~second~~ volume of *The analysis of natural waters* covers direct preconcentration techniques (i.e. those not involving complex formation). These include adsorption of the substance to be preconcentrated on solid materials such as resins, polymers, metal oxides, metals, and carbon followed by desorption with a small volume of a suitable desorbent and suitable analysis. Also included are techniques based on coprecipitation of the species to be preconcentrated with another species, direct solvent extraction techniques, head space and purge and trap techniques, and a variety of other techniques including electrolysis, gas evolution, flotation, evaporation/distillation, precipitation/cocrystallization, dialysis, osmosis, and freeze drying.

As with the first volume, the emphasis is laid on providing practical experimental detail so that the reader can, in many cases, apply the methods without reference to source literature and will be in a position to adapt procedures to his or her particular requirements.

Anglesey T. R. C.
July 1992

PREFACE TO VOLUME 1

Despite the great strides forward in analytical instrumentation that have been made in the last decade, the analyst working in the fields of potable water analysis and environmental analysis of non-saline and saline waters finds that, frequently, the equipment has insufficient sensitivity to be able to detect the low concentrations of organic and inorganic substances present in his samples with the consequence that he has to report less than the detection limit of the method. Consequently, trends upwards or downwards in the levels of background concentrations of these substances in the environment cannot be followed. This is a very unsatisfactory situation which is being made worse by the extremely low detection limits being set in new directives on levels of pollution, issued by the European Community and other international bodies. To overcome the problem, there has been a move in recent years, to apply preconcentration to the sample prior to analysis so that, effectively, the detection limit of the method is considerably reduced to the point that actual results can be reported and trends followed.

The principle of preconcentration is quite simple. Suppose that we need to determine 5 ng litre^{-1} of a substance in a sample and that the best technique has a detection limit of 1 μg litre^{-1} (1000 ng litre^{-1}). To reduce the detection limit to 5 ng litre^{-1}, we might, for example, pass 1 litre of the sample down a small column of a substance that absorbs the substance with 100 per cent efficiency. We would then pass down the column 5 ml of a solvent or reagent which completely dissolves the substance from the column thereby achieving a preconcentration of $1000/5 = 200$. Thus, if the detection limit of the analytical method without preconcentration were 1000 ng litre^{-1}, then with preconcentration it would be reduced to approximately 5 ng litre^{-1}.

The use of a column is but one of many possible methods of achieving preconcentration. Each chapter of the book discusses a different method of preconcentration and its application to the preconcentration of cations, anions, organic substances, and organometallic compounds. The book is based on a survey of the recent world literature dealing with preconcentration and its application to saline and non-saline waters.

A combination of preconcentration with the newest, most sensitive, and, by definition, most expensive analytical techniques now becoming available is achieving previously undreamt of detection limits at the very time when the requirements for such sensitive analysis is increasing at a rapid pace. Thus the combination of preconcentration with graphite furnace, Zeeman or inductively coupled plasma atomic absorption spectrometry and, particularly, the combination of the latter technique with mass spectrometry is enabling exceedingly low concentrations of metals in the ng $litre^{-1}$ or lower range to be determined. Preconcentration prior to gas or high performance liquid chromatography is achieving similar results in the analysis of organics.

Another aspect of preconcentration is, however, worthy of mention, particularly in the case of smaller laboratories which cannot afford to purchase the full range of modern analytical instrumentation. Using older, less sensitive instrumentation preconcentration will still achieve very useful reductions in detection limits which will be adequate in many but not all instances. Thus, if conventional atomic absorption spectrometry achieves detection limits of 1 and 5 mg $litre^{-1}$ for cadmium and lead in water, then a 200-fold preconcentration will reduce these limits to approximately 5 and 25 μg $litre^{-1}$ and a 1000-fold preconcentration will achieve 1 and 5 μg $litre^{-1}$.

The first volume of *The analysis of natural waters* covers those techniques which involve the formation of a complex of the species to be preconcentrated followed by concentration of the complex by techniques based on solvent extraction or adsorption on a column. This is followed by desorption with a small volume of a suitable reagent and then, in each case, by analysis using an appropriate analytical technique.

Throughout the book emphasis is laid on providing practical experimental detail so that the reader can, in many cases, apply the methods without reference to source literature and will be in a position to adopt procedures to his or her particular requirements. This is a field where much remains to be discovered and it is hoped that this book will assist chemists to further develop procedures.

Whilst the book has been written with the interest of water chemists in mind, many of the preconcentration procedures discussed could with little or no modification be applied to improving detection limits in laboratories in a wide range of other industries including, electronics, semiconductors, metallurgy, pharmaceuticals, organic chemicals, petroleum, and polymers.

The following groups of people will find much to interest them: management and scientists in all aspects of the water and other industries, river management, fishery industries, sewage effluent treatment and disposal, and land drainage and water supply, as well as management and scientists in all branches of industry which produce aqueous effluents. It will also be of interest to agricultural chemists; agriculturalists concerned with the ways in which chemicals used in crop and soil treatment permeate the ecosystem; to the biologists and scientists involved in fish, insect, and plant life; and to the medical profession, toxicologists, public health workers, and public analysts. Other groups of workers to whom the book will be of interest include oceanographers, fisheries experts, environmentalists, and, not least, members of the public who are concerned with the protection of the environment. The book will also be of interest to practising analysts, and, not least, to the scientists and environmentalists of the future who are currently passing through the university system and on whom, more than ever previously, will rest the responsibility of ensuring that by the turn of the century we are left with a worthwhile environment to protect.

Anglesey T.R.C.
July 1992

CONTENTS

List of contents - Volume 1 xi

1 Solid organic adsorbents and chelators 1
- 1.1 Metal cations 1
- 1.2 Anions—non-saline waters 39
- 1.3 Organics 40
- 1.4 Organometallic compounds 52
- References 52

2 Metal oxides and metal adsorbents 60
- 2.1 Metal cations 60
- 2.2 Anions 68
- 2.3 Organics 69
- 2.4 Organometallic compounds 73
- References 74

3 Active carbon adsorbent 77
- 3.1 Metal cations 77
- 3.2 Anions 81
- 3.3 Organics 81
- References 89

4 Inorganic and organic coprecipitation techniques 92
- 4.1 Inorganic coprecipitants—introduction 92
- 4.2 Inorganic coprecipitants—metal cations 92
- 4.3 Inorganic coprecipitants—anions 110
- 4.4 Inorganic coprecipitants—organics 111
- 4.5 Organic coprecipitants—introduction 112
- 4.6 Organic coprecipitants—metal cations 113
- References 119

5 Solvent extraction of organic and organometallic compounds 124
- 5.1 Organic compounds—non-saline waters 124
- 5.2 Organic compounds—sea-water 147
- 5.3 Organometallic compounds 151
- References 152

6 Purge and trap and head-space analysis techniques 162
- 6.1 Purge and trap techniques 162
- 6.2 Head-space analysis 179
- References 189

7 Miscellaneous preconcentration techniques 195
7.1 Metal cations 195
7.2 Anions 225
7.3 Organics 225
7.4 Organometallic compounds 235
References 235

Index 241

CONTENTS OF VOLUME 1

1 INTRODUCTION

2 CHELATION–SOLVENT EXTRACTION OF ANIONS AND CATIONS

3 MACRORETICULAR NON-POLAR RESINS

4 CATION EXCHANGE RESINS

5 ANION EXCHANGE RESINS

1

SOLID ORGANIC ADSORBENTS AND CHELATORS

1.1 Metal cations

1.1.1 Chitosan

Non-saline waters Chitosan (deacetylated chitin) has been used for the preconcentration of molybdenum[1-5] and of zirconium, niobium, cerium, and ruthenium[6] in nuclear fuel solutions. Jha *et al.*[7] the investigated the efficiency of removal of cadmium from water by chitosan.

Sea-water Muzzarelli and Rocchetti[8] showed that chitosan is superior to Dowex A700 ion exchange resin and modified celluloses for the preconcentration of copper from unoxidized sea-water. In this procedure the sample is passed through a column (30 × 3 mm) packed with chitosan (100 mg: 100–200 mesh) and the chelated chopper eluted with a 1 per cent solution of 1,10-phenanthroline (20 ml) at 50 °C or with 1 M sulphuric acid (20 ml). An aliquot (20 μl) in place in a hot graphite analyser programmed to dry for 20 s, char for 20 s, and atomize for 10 s. The amount of copper present in determind by comparison with standards. Average recoveries from the column were 100 per cent and the coefficient of variation was 7.5 per cent for 3.4 μg of copper per litre.

Muzzarelli and Sipos[9] showed that a column of chitosan (15 × 10 mm) can be used to concentrate zinc from 3 litres sea-water before determination by anodic stripping voltammetry with a composite mercury–graphite electrode. Zinc (and lead) are eluted from the column by 50 ml 2 M ammonium acetate, copper by 10 ml of 0.01 M EDTA, and cadmium by 3 ml 0.1 M potassiun cyanide. Other workers[10,11] have used chitosan for the preconcentrations of cations from sea-water.

1.1.2 Silica and glass bead immobilized chelators

Metal chelating resins and immobilized (adsorbed or chemically bonded) chelates have found widespread application for the concentration and/or separation of trace metals from a variety of matrices. Dimethylglyoxime, alkylamines, diamines, xanthates, and dithiocarbamates, propylenediaminetetra-acetic acids, *n*-butylamides, *N*-substituted hydroxylamine, hexylthioglycolate, ferroin-type chelating agents, iminodiacetate, amidoxime, dithizone, and 8-hydroxyquinoline have all been immobilized on various substrates. Tailoring chemically bonded chelating agents to specific needs allows the use of selective or general concentration schemes and also permits the 'recycling' of the chelating agent. Their major use lies in the preconcentration of trace metal ions from aqueous and saline media.

Non-saline waters Various metal preconcentration systems have been devised based on the immobilization of organic complexing agents on silica gel. These organic reagents include 8-quinolinol,[12] 2,2′-dipyridyl-4-amino-3-hydrazino-5-mercapto-1,2,4-triazole hydazone,[13,14] dithizone, and diphenylcarbazone.[16]

Marshall and Mottola[12] used silica immobilized 8-quinolinol to preconcentrate trace metals on-line prior to their determination by atomic absorption spectrometry. Samara and Kaouimtzis[14] used 2,2′-dipyridyl-4-amino-3-hydrazino-5-mercapto-1,2,4-triazole hydrazone supported on silica gel to preconcentrate metal ions (copper, lead, nickel, zinc, cadmium and cobalt) from waters prior to their determination by atomic absorption spectrometry. Genova *et al.*[15] preconcentrated cadmium on dithizone supported on silica gel and extended the detection range down to 10 μg.

Copper, lead, cadmium and zinc ions in water have been preconcentrated with 2-mercaptobenzthiazole loaded on to controlled pore size glass beads with the aid of collodion.[17] Cadmium and zinc were not retained quantitatively even at low flow rates. Selective preconcentration of copper and lead was therefore possible by passing the sample through the column at a high rate at pH 6.5. The effects of cobalt, iron, nickel, EDTA, citric acid, and tartaric acid on the process were examined. The copper and lead retained on the column were eluted together with the collodion with methyl isobutyl ketone by batch mode elution and determined by one-drop atomic absorption spectrometry.

Laydon *et al.*[18] used sodium diethyldithiocarbamate immobilized on controlled pore size glass beads for the preconcentration of silver, arsenic, cobalt, chromium, copper, iron, mercury, manganese, lead, and zinc. They compared the results obtained by this procedure with methods based on direct precipitation of the metals with organic complexing agents including sodium diethyldithiocarbamate, ammonium pyrrolidinedithiocarbamate, and oxine. The conclusion reached in this work was that much more experimental work was required to test method reliability.

Allen *et al.*[19] compared the efficiencies of 8-hydroxyquinoline and EDTA immobilized on controlled pore size glass for the preliminary concentration of traces of aluminium in natural waters, prior to determination by flame atomic absorption spectrometry. Both chemicals were satisfactory at sample pH values greater than 4.6. Szabo *et al.*[20] used 8-hydroxyquinoline-substituted silica gel to preconcentrate yttrium-90 from solutions containing strontium-90. Yttrium was desorbed from the column with hydrochloric acid and converted to the oxalate prior to counting in a low background anti-coincidence beta-counter.

Sea-water Diphenylcarbazone and diphenylcarbazide have been widely used for the spectrophotometric determination of chromium.[21] Only relatively recently, however, has the nature of the complexation reactions been elucidated. Cr^{III} reacts with diphenylcarbazone where Cr^{VI} reacts (probably via a redox reaction combined with complexation) with diphenylcarbazide.[22] Although speciation would seem a likely prospect with such reactions, commercial diphenylcarbazone is a complex mixture of several components, including diphenylcarbazide, diphenylcarbazone, phenylsemicarbazide, and diphenylcarbadiazone with no stoichiometric relationship between the diphenylcarbazone and diphenylcarbazide.[23] As a consequence, use of diphenylcarbazone to chelate Cr^{III} selectively also results in the sequestration of some Cr^{VI}. Total chromium can be determined with diphenylcarbazone following reduction of all chromium to Cr^{III}.

Use of immobilized chelating agents for sequestering trace metals from aqueous and saline media presents several significant advantages over chelation – solvent extraction approaches to this problem.[24,25] With little sample manipulation, large preconcentration factors can generally be realized in relatively short times with low analytical blanks. As a consequence of these considerations Willie *et al.*[26] developed a new approach to the determination of total chromium. This involves preliminary concentration of dissolved chromium from sea-water by means of an immobilized diphenylcarbazone chelating agent, prior to determination by atomic absorption spectrometry. A Perkin-Elmer Model 500 atomic absorption spectrometer fitted with a HGA-500 furnace with Zeeman background correction capability was used for chromium determinations. Chromium was first reduced to Cr^{III} by addition of 0.5 ml aqueous sulphur dioxide and allowing the sample to stand for several minutes. Aliquots of sea-water were then adjusted to pH 9.0 ± 0.2 by using high-purity ammonium hydroxide and gravity fed through a column of silica at a nominal flow rate of 10 ml min^{-1}.

The sequestered chromium was then eluted from the column with 10.0 ml 0.2 M nitric acid. More than 93 per cent of chromium was recovered in the first 5 ml of eluate by this method. Extraction of 80 ng spikes of Cr^{III} from 200 ml aliquots of sea-water was quantitative. Neither Cr^{III} nor Cr^{VI} could be quantitatively extracted. Results for the analysis of a near-shore sample of sea-water and open ocean trace metal reference sea-water, NASS-1 are given in Table 1.

Sturgeon *et al.*[27] preconcentrated cadmium, copper, zinc, lead, iron,

Table 1 Concentration analysis (μg litre^{-1}) of sea-water for total chromium.

Trial	Coastal water (salinity = 29.5%)	Open-ocean. NASS-1 (salinity = 35.0%)
1	0.100	0.19
2	0.096	0.15
3	0.095	0.18
4		0.19
Mean (±SD)	0.097 ± 0.003	0.18 ± 0.02
Accepted value	0.10 ± 0.01	0.184 ± 0.016

From Willie *et al.*[26] with permission.

manganese, nickel, and cobalt from sea-water on to silica immobilized 8-hydroxyquinoline prior to determination by graphite furnace atomic absorption spectrometry. Results for the analyses of a near-shore sea-water sample are given in Table 2. Near-shore samples were concentrated 50-fold, the open-ocean samples 90-fold. Calibration was achieved by spiking an aliquot of the concentrate with the element of interest, thereby obtaining an exact matrix match. Results obtained by using the immobilized 8-hydroxyquinoline concentration procedure were compared to 'accepted' values for these samples. Good agreement with accepted values is evident for all three samples.

p-Dimethylaminobenzylidinerhodamine and 5,7-dibromo-8-hydroxyquinoline[6] on silica gel have been used for the adsorption of metals from sea-water.[28] Volkan *et al.*[29] described a procedure for the preconcentration of cadmium, copper, lead, and zinc from sea-water by

Table 2 Analyses of near-shore sea-water (salinity 29.5‰) (means ± SD; n = 3).

Element	Concentration (μg ml^{-1})			
	Sample 1		Sample 2	
	I-8-HOQ	Accepted value	I-8-HOQ	Accepted value
Cd	0.020 ± 0.001	0.024 ± 0.004	0.025 ± 0.001	0.023 ± 0.001
Pb	0.22 ± 0.01	0.22 ± 0.06	0.014 ± 0.003	0.018 ± 0.001
Zn	0.44 ± 0.01	0.41 ± 0.05	0.29 ± 0.03	0.28 ± 0.01
Cu	1.03 ± 0.06	0.96 ± 0.04	0.17 ± 0.01	0.22 ± 0.02
Fe	1.0 ± 0.1	1.03 ± 0.04	7.2 ± 0.9	6.9 ± 0.02
Mn	0.71 ± 0.02	0.68 ± 0.05	1.06 ± 0.01	1.13 ± 0.02
Ni	0.33 ± 0.01	0.31 ± 0.04	0.39 ± 0.01	0.34 ± 0.01
Co	0.018 ± 0.002	0.015 ± 0.007	0.017 ± 0.001	ND

ND = Not determined.
From Sturgeon *et al.*[27] with permission.

Table 3 Results of copper determination samples (Environmental Protection Agency.

Sample	Cu concentration (ng ml^{-1})	
	Determined	Reported
EPA-1	51 ± 4.6	50
EPA-2	252 ± 2.6	250
EPA-3	38 ± 3.0	40
EPA-4	8.9 ± 1.6	8.3
Tap water	26 ± 2.0	NA

Other elements known to be present (concentration in ng ml^{-1} in parentheses); EPA-1, Al (450), As (60), Be (250), Cd (13), Cr (80), Co (80), Fe (80), Pb (120), Mn (75), Hg (3.5), Ni (80), Se (30), V (250), Zn (80); EPA-2, Al (700), As (200), Be (750), Cd (50), Cr (150). Co (500), Fe (600), Pb (250), Mn (350), Hg (7.5), Ni (250), Se (40), V (750), Zn (200); EPA-3, Al (350), As (40), Be (150), Cd (10), Cr (60), Co (70), Fe (50), Pb (80), Mn (55), Hg (3.0), Ni (70), Se (20), V (200), Zn (60); EPA-4, Al (106), As (27), Be (29), Cd (9.1), Cr (7.1), Co (43), Fe (22), Pb (43), Mn (13), Hg (0.7), Ni (17), Se (11); tap water, unknown. Stillwater, OK.
NA = Not available.
From Marshall and Mottola[12] with permission.

adsorption on silica gel modified by treatment with (3-mercaptopropyl) trimethyloxysilane. Nakashima *et al.*[30] preconcentrated cadmium, lead, zinc, copper, iron, manganese, nickel, and cobalt from sea-water on to a column of silica immobilized by 8-hydroxyquinoline prior to analysis by graphite furnace atomic absorption spectrometry. Detection limits ranged from 0.2 μg $litre^{-1}$ (cobalt) to 40 μg $litre^{-1}$ (iron) based on 50 ml sample volumes.

Marshal and Mottola[12] have used silica immobilized 8-quinolol as a means of preconcentrating metals for analysis by flow injection atomic absorption spectrometry. This has proved to be a particularly useful material for sample preparation, matrix isolation, and preconcentration of trace metal ions.[31–35] The selectivity of silica immobilized 8-quinolol for transition metal ions over the alkali and alkaline earth metal ions makes it useful for samples containing large quantities of the latter such as sea-water. These workers evaluated breakthrough capacities under different flow, temperature, and geometric characteristics of the preconcentration column. The columns have relatively high capacities for metals and do not suffer from complications due to swelling. Excellent agreement was obtained in determinations of copper on standard enviromental samples (Table 3).

1.1.3 Chemically modified polymers—non-saline waters

Aminodiacetate resins, Chelex-100 The chelating resin Chelex-100, a purified form of the strongly acidic Dowex Al cation exchange resin, has been increasingly used in recent years for the separation and preconcentration of trace metals from natural waters. It is particularly suited to this application because optimum metal removal occurs in the pH range of natural waters, pH 6–8, where distribution coefficients of the order of 10^5 have been measured.[31]

In most instances, Chelex-100 has been used in the H^+ form, although the effluent pH under these conditions may be as low as 2.8. Florence and Batley[32] showed that the H^+ form did not completely remove labile zinc, cadmium, lead, and copper from sea-water until the passage of more than 500 ml of sample had increased the pH of the effluent to 6.5. Treatment of the column with sodium acetate before use raised the effluent pH to 7.1. The ammonium or calcium forms also have been used to circumvent this problem.[32–35] Chelex-100 is also used in the sodium form.

Several papers have reported distribution data for selected metal ions on Chelex-100 as a function of pH.[35,36] Batch equilibration studies reported by Dow Chemical Company[37] show complete removal of cobalt, nickel, and copper at pH 4, while similar results have been obtained for zinc, cadmium, and manganese and for iron(III) and lead.[38] Exchange kinetics are slow at low pH values where the imino nitrogen of the resin functional group is protonated, however, at the pH of natural waters, exchange is rapid and efficient column operation is observed in a 1.5 ml column with flow rates of 1.5–2.5 ml min^{-1}.

Vautsa *et al.*[39] compared three different chelating resins, Chelex-100, Hyphan cellulose and Amberlite IRC 718 with respect to their preconcentration efficiency for traces of copper from fresh-waters. Interferences by some complexing agents such as EDTA, NTA, sodium tripolyphosphate, and humic acids were also investigated. The retention of copper on the three chelating resins was quantitative above pH 3. The complete elution of copper from Chelex-100 and Hyphan resins was only achieved by a mixture of nitric and hydrochloric acids. None of the eluting solutions tested completely eluted copper from Amberlite.

Atomic absorption spectrometry Pakalns *et al.*[40,41] have described the use of Chelex-100 for the preconcentration of metals from natural and polluted non-saline waters. In particular, they looked at the effect of some organic surface active pollutants on the efficiency of metal removal. These pollutants include cationic, anionic and non-ionic detergents, formulated detergents, detergent additives, and a soap, which are all likely domestic and industrial discharges into river systems. These workers showed that trace amounts of zinc, cadmium, copper, nickel, manganese, cobalt and lead could be separated from natural waters on Chelex-100 resin (50–100 mesh) in the presence of all the aforementioned pollutants.

Metal recoveries are better than 92 per cent but are poor in the presence of soap or the detergent additive, nitrilotriacetic acid. Although strong adsorption of cationic and to a lesser extents anionic and non-ionic detergents, occurs on the resin surface, low recoveries can be attributed to incomplete metal elution rather than to blockage of adsorption sites. Total metal present in natural waters is not adsorbed by Chelex-100 unless metal ions are first released from colloids or strong complexes. Destruction of the complexes by ultraviolet light or an acid digestion before the sample is applied to the Chelex-100 column reaults in a complete recovery of metals.

The efficiency of Chelex-100 in concentrating metals in the presence of surfactants and the subsequent effectiveness of the eluant were studied with radiotracers. The tracer solutions ^{54}Mn, ^{65}Zn, ^{60}Co, and ^{115}Cd were equilibrated with the corresponding standard metal solutions, surfactant was added and the recommended procedure followed. Effluent and eluate fractions were collected and the distribution of the metals determined by gross γ-counting in a well-crystal γ-counter. Blocking of adsorption sites by surfactants did not occur and metal adsorption was very good in the presence of the surfactants studied. Low metal recoveries could in general, be attributed not to initial blockage of adsorption sites but to the incomplete removal of metal from the column with the acid eluant. This could result from metal being trapped within the resin structure as the volume of the resin decreased (normally twofold) on elution with acid. During this step, detergents, particularly the anionic and non-ionic species, could hinder elution by partially blocking the resin pores. In the presence of nitriloacetic acid, incomplete metal tracer adsorption was observed. Similar results were obtained in the presence of humic acids.

To gain insight into the fraction of total metal determined in natural waters after concentration by Chelex-100, small quantities of manganese-54, zinc-65, cobalt-60, and cadmium-115 traces were mixed with four different filtered natural water samples. These were allowed to stand for up to seven days before the mixtures were put through the resin column. The proportion of metal tracer adsorbed by the resin decreased with standing time. Because up to 28 per cent of tracer was not recoverable by the resin after storage for seven days, the metal concentrations could not be regarded as total metal concentrations. Moreover, any metal present either as strong complexes which do not (or only slowly) exchange with the iminodiacetate resin groups, or as metal associated with colloidal particles, would not be retained by the chelating resin. Colloidal particles cannot penetrate the pore network of Chelex-100 which has an average pore diameter of only 6 nm.

Subramanian *et al.*[42] have discussed on-site sampling with preconcentration for the determination of some Chelex-100 levels of labile metals (cadmium, copper, lead, and zinc) in potable water. The on-site pump integrated water sampler coupled with a Chelex-100 preconcentration unit is described in detail. Metals in the concentrate were determined by graphite furnace atomic absorption spectrophotometry.

The Chelex-100 column used in conjuction with the pump integrated and batch integrated samplers are prepared for ion exchange as follows. Place a small amount of precleaned glass wool at the bottom of the column. Slurry load about 2 g of the resin into the column allowing the water to drip out of the column as the slurry settles. Pass two 20 ml portions of 3 M nitric acid through the column followed by two 20 ml portions of high purity water and 10 ml of 2 M ammonia. Wash the column with high purity water until the effluent pH is near neutral. Incorporate the column into the pump integrated sampler preconcentration assembly.

On completion of exchange disconnect the column, rinse once with 10 ml high purity water, thrice with 10 ml aliquots of 1 M ammonium acetate buffer (pH 5.2), and finally with two 10 ml portions of high purity water. Transfer the resin and glass wool from the column into a 10 ml polyethylene conical flask and equilibrate for 1 h with 30 ml of 3 M nitric acid. Vacuum filter the contents of the flask through a Pyrex glass frit of medium porosity. Collect the filtrate in a 50 ml Pyrex glass volumetric flask, and wash the glass frit thrice with small jets of water. Bring the flask to the 50 ml mark with high purity water.

For the determination of a given metal, withdraw a 20 μl aliquot from the volumetric flask with an Eppendorf pipette fitted with disposable plastic tips and inject into the pyrocoated graphite tube mounted in the graphite furnace of the atomic absorption spectrometer. Prior to use the tips were decontaminated from traces of the metals by soaking them for 24 h in 1 per cent nitric acid followed by four rinses with high purity water. The sequential dry–char–atomize program of the Perkin Elmer HG A500 graphite furnace atomic absorption spectrometer given in Table 4 was followed and the peak absorbence noted. The amount of each metal in the potable water sample was obtained by the method of standard addition.

Subramanian *et al.*[42] confirm the conclusions reached by Pakalns *et al.*,[40] namely that Chelex-100 will remove only that fraction of the trace metal level that is 'Chelex labile' and founds for example, that only 44–63 per cent of lead and copper in potable water is labile, presumably due to the presence of humic acid and possibly other chelators in the water. They do not discuss decomplexing using ultraviolet light or acid digestion methods.

Chelex-100 resins have also been used for the preconcentration of molybdenum;[43] zinc, cadmium, mercury and lead;[44] iron, copper, nickel, cadmium, cobalt, zinc, lead, and manganese;[45] and uranium.[46]

Table 4 Optimized GFAA parameters for Cd, Cu, Pb, and Zn.[a]

Parameter	Cadmium	Copper	Lead	Zinc
Wavelength (nm)	228.8	324.8	383.3	213.9
Slit (nm)	0.7	0.7	0.7	0.7
Drying temperature (°C)	100	100	100	100
Drying time, ramp-hold (s)	20–20	20–20	20–20	20–20
Ashing temperature (°C)	250	600	350	400
Ashing time, ramp-hold (s)	20–20	10–30	20–20	20–20
Atomization temperature (°C)	1900	2400	2100	2000
Atomization time, ramp-hold (s)	1–4	1–4	1–4	1–4

[a]Perkin-Elmer Model 603 AA spectrometer equipped with the HGA-500 graphic furnace and deuterium arc background corrector.
From Subramanian *et al.*[42] with permission.

Neutron activation analysis. Gladney *et al.*[47] preconcentrated uranium-235 and uranium-238 in natural water on Chelex-100 resin and subsequently measured the concentration of these elements in the resin by neutron activation analysis. Hinse *et al.*[48] and Greenberg and Kingston[49] have also used neutron activation analysis to determine metals preconcentrated from natural waters on to cation exchange resins.

X-Ray fluorescence spectrometry. This technique has also been used in the analysis of preconcentrated cations.[50,51] Thus, Van Espen *et al.*[51] preconcentrated toxic metals in potable water on Chelex-100 filters and determined the elements directly on the filters in amounts down to 1 μg litre^{-1} by energy dispersive X-ray flouresence spectrometry.

Anodic scaming voltammetry. Figura and McDuffie[52] preconcentrated copper, cadmium, lead, and zinc by adsorption on Chelex-100 resin, enabling sub-μg litre^{-1} quantities of these elements to be determined. The Chelex column effluents were analysed by anodic scanning voltammetry.

Sephadex gel Molybdenum has been determined after preconcentration on Sephadex G. 25 gel at pH 3.5. Ethylenediaminetetra-acetic acid desorbs molybdenum from the gel. This preconcentration method brings the detection limit for molybdenum in fresh and sea waters down to 1μg litre^{-1} in 250-ml samples. The procedure was applied to 100-ml solutions containing 1.25 μmol of molybdenum and 1- to 10000-fold amounts of various ions. Interference from vanadium(V) and tungsten(VI) is probably due to complex formation with molybdenum. Iron(III) at a 100-fold level lowered the recovery, probably because its hydroxides or hydroxo complexes adsorb molybdenum, but the interference could be overcome in the presence of 0.5 M acetate. The method is suitable for the determination of molybdenum at the levels normally encountered in river water (0.2–0.6 μg litre^{-1}) and is particularly effective for sea-water.

Vinyl dithiocarbamate resins Minagawa *et al.*[53] have described a technique employing chelating resins which has been applied to the determination of very low concentrations (down to 0.2 ng litre^{-1}) of organic and inorganic mercury in natural waters including rivers, lakes, and rain waters. The resin used contains dithiocarbamate groups which bind mercury but not alkali and alkaline earth metals. Both forms of mercury are collected at pH 1–11 and eluted with slightly acidic 4 per cent thiourea in water. Large volumes of water can be concentrated to determine mercury by cold vapour atomic absorption spectrometry. Mercury vapour is generated from inorganic mercury with alkaline stannous chloride and from organic and inorganic mercury with a cadmium chloride–stannous chloride solution.

The calibration graphs obtained for mercury(II) and CH_3Hg^+ were identical and are linear over the range 0–12 ng. Possible interferences are other ions, amino acids, and naturally occurring chelating agents which could affect the preconcentration, desorption, and reduction steps. No interference was produced in the determination of 0.1 μg of mercury(II) by the presence of at least 1000 μg of each of the following ions or substances added to 5-litre 1 aliquots of river water: chromium(III), magnesium, sodium, potassium, calcium, nickel, copper(II), lead(II), cadmium, gold(III), iron(III), aluminium, zinc, phosphate, chloride, carbonate, nitrate, sulphate, silicate, cysteine, and humic acid. The accuracy of the method was tested by analysing river water samples spiked with known amounts of mercury and CH_3Hg^+. The accuracy of the method was satisfactory. The precision of the method was 9.5 ± 0.43 ng litre^{-1}.

Yamagani *et al.*[54] also applied chelating resins (dithiocarbamate type) to the determination of μg litre^{-1} levels of mercury in water. The samples are adjusted to pH 2.3 and passed through a column packed with 5 g of the resin, at a flow rate of 50 ml min^{-1}. The resin is then digested under reflux with concentrated nitric acid and the mercury is determined by atomic absorption spectrophotometry, using the reduction aeration technique. The method is relatively simple and inexpensive and the detection limit is 10 ng mercury in water samples as large as 10 litres.

Nicotinium molybdophosphate resin When metal solution adjusted to pH 2.5, were passed through ion exchange columns containing nicotinium molybdophosphate resin sodium, potassium, calcium, and magnesium were selectively eluted using 4 m ammonia solution. The recovered ions were analysed by atomic absorption spectrophotometry. These procedures achieved recoveries of 98–102 per cent for cadmium, chromium, manganese, nickel, lead, zinc, cobalt, iron, and copper. Recovery was practically independent of the initial heavy metal concentration (5–10 mg litre^{-1}).

Poly(triaminophenyl)glyoxal This polymer has been used as a chelating agent for the determination of traces of nickel in sea-water, natural water, and trade effluent.[55] Nickel is desorbed from the polymer with 0.1 M hydrochloric acid, extracted into an isobutyl methyl ketone solution of ammonium pyrrolidinedithiocarbamate and determined by atomic absorption spectrophotometry at 232 nm.

Thioamide substituted polyacrylonitrile Samchuk *et al.*[56] preconcentrated copper, lead, cadmium, nickel, cobalt, and zinc from waste water on fibrous thioamide substituted polyacrylonitrile, prior to desorption and determination by atomic absorption spectrometry.

Cellulose derivatives

Cellulose piperazine dithiocarboxylate. Imai *et al.*[57] preconcentrated 18 trace elements by passing the natural water sample through a column packed with this reagent. The solid was then ashed and elements determined by neutron activation analysis. This procedure was applied to fresh-water samples.

2,2′-Diaminodiethylaminecellulose. This material $\left(\text{cellulose}-N\begin{smallmatrix} \diagup CH_2CH_2NH \\ \diagdown CH_2CH_2NH_2 \end{smallmatrix}\right)$ has been examined by several workers[58,59] for the pre-concentration of transition elements in natural and potable waters prior to analysis by X-ray flourescence spectrometry[58,59] and neutron activation analysis.[39]

Smits and Van Grieken[58] showed that a 2,2′-diaminodiethylamine-cellulose filter is a very simple and effective method for preconcentrating metals prior to X-ray fluorescence spectromentry and it is an ideal target material in this technique. Detection limits are around 0.5 μg litre^{-1}.

X-Ray fluorescence spectrometer:

Energy dispersive XRF unit including a high voltage generator and Kevex-0810 subsystem with W-anode X-ray tube, secondary targets of Ti, Ge, Mo, Ag, Sn, and Nd and a turnable 16 sample holder. The X-rays were registered by a 30 mm^2 Si(Li) detector with 170 eV FWHM at 5.9 keV, amplification and pulse pile up rejection systems and a 4096-channel analyser and magnetic tape recording unit for off-line computer spectrum evaluation. For calibration a series of commercial thin-film standards was available.

Preparation of filter:

Soak dried Whatman-41 cellulose filters (5.5 cm diameter) for 30–60 min in 200 ml dry *N,N*-dimethylformamide (distilled with 10 ml benzene/100 ml dimethylformamide). After adding 6 ml phosphorus oxychloride, heat the mixture to 90 °C for at least 15 min. Wash the filters successively with dimethylformamide, water, 5 per cent sodium hydroxide, 5 per cent acetic, acid, and water. Heat the filters with stirring, in an excess of purified diamine diethylamine at 130 °C for 2 h.

Filtration unit:

A plexiglass filtration unit of straightforward cylindrical shape was used to yield a homogeneous flow over a well defined filter area of 10.2 cm^2. Collection of suspended matter, if required, was accomplished previously in a separate filtration over a nuclepore membrane.

Procedure:

Filter water samples (100 ml) at natural pH separately through a Nuclepore membrane and then, at a rate of less than 1 ml min^{-1} cm^{-2},

Table 5 Interference-free detection limits for some elements (collected as cations or anions) for 3000 s XRF-measurement after preconcentration on a DEN-filter.

Element	Detection limits (μg $litre^{-1}$) for a 1 litre water sample collected on a 10-cm^2 DEN-filter			
	Secondary fluorescer system			
	Ge	Mo	Ag	Sn
V	0.07	0.25	0.8	0.9
Cr	0.07	0.2	0.6	0.8
Mn	0.05	0.2	1.4	0.8
Fe	2.2	2.5	2.5	3.3
Cu		0.2	0.8	1.0
Zn		0.4	0.4	0.4
As		0.1	0.3	0.2
Se		0.05	0.1	0.1
La[a]		1.4	3.5	4.9
W[a]		0.35	0.3	0.5
Hg[a]		0.3	0.25	0.4
Pb[a]		0.2	0.4	0.2
U[a]		0.2	0.2	0.3

[a] Detection through *L*-lines, *K*-lines in other cases.
From Smits and Von Grieken[58] with permission.

Table 6 Average apparent collection yield and precision for 16 analyses of synthetic water samples preconcentrated onto DEN-filters.

Ion	Apparent collection yield (per cent)	Relative standard deviation per measurement (per cent)
Cr^{3+}	104	9
Fe^{3+}	104	10
Co^{2+}	97	10
Ni^{2+}	90	14
Cu^{2+}	100	11
Zn^{2+}	105	11
Eu^{3+}	105	9
Pb^{2+}	102	8
UO_2^{2+}	97	7

From Smits and Van Grieken[58] with permission.

through a 2,2′-diaminodiethylaminecellulose filter. After drying in a desiccator suspend the filters on a Teflon ring fitting into the XRF unit sample holder and analyse for 3000 s, e.g. using molybdenum secondary fluorescent excitation at 40 kV, 40 mA tube conditions.

In Table 5 are listed some typical detection limits of the system (three times the square root of the background count rate depending on scatter, on the characteristic peaks from blank filter impurities, and proportional to the filter thickness). These hypothetical detection limits assume no mutual interference of the elements; in practical cases the detection limits can be somewhat less favourable. Most elements can be measured down to 0.5 μg litre^{-1}. Table 6 lists the average apparent collection yields, calculated relative to commercially available thin film standards after X-ray absorption correction. Except for nickel(II) the apparent positive and negative deviations from a 100 per cent collection yield can all be attributed to uncertainties in the X-ray fluorescence analysis and calibration procedure. Both the accuracy and precision of the overall procedure appear to be around 10 per cent.

In Table 7 the generally good agreement results obtained by this procedure are compared with those obtained by two other more laborious procedures. The results from the activated carbon method appear slightly higher, probably because this procedure includes the trace metals in colloids and strong complexes with humic material.

In more recent work Reggeks and Van Grieken[59] showed that 2,2′-diaminodiethylaminecellulose powder offered better chelation capacity to the filters discussed above. The powder was insensitive to high concentrations of alkali and alkaline earth ions and to humic substances. Zinc, cadmium, and manganese could be efficiently eluted from powders

Table 7 Transition metal concentration in tap-water from drinking water supply, measured by XRF in combination with different preconcentration techniques.

Element	Concentration (and stand. dev. per meas.) (μg. litre^{-1})			
	Suspended collected on 0,4 μm Nuclepore membrane	Dissolved, determined after preconcentration using		
		2,2′-diethylaminodiethylcellulose filters	1-(-2-pyridylazonaphthol-cocrystallization	Oxine chelation and activated carbon adsorption
Mn	0.4(0.1)	–	0.4(0.2)	0.9(0.1)
Fe	2.8(0.6)	2.5(0.3)	3.4(0.5)	4.5(0.3)
Ni	<0.2	2.3(0.1)	2.3(0.2)	2.4(0.1)
Cu	0.1(0.02)	1.8(0.1)	1.3(0.1)	2.2(0.1)
Zn	0.1(0.04)	2.7(0.5)	2.3(0.2)	3.5(0.2)
Pb	<0.2	<0.1	<0.1	–

From Smits and Van Grieken[58] with permission.

for subsequent analysis, by X-ray fluorescence spectromentry or alternative techniques such as neutron activation analysis and atomic absorption spectrometry. Cellulose powder with 2,2′-diaminodiethylamine functional groups exhibits efficient complexation of transition metal cations. Collection yields above 85 per cent are obtained up to a chelation capacity of 1.5 meq g^{-1}. Since good recovery of metals is obtained for a pH greater than 5, no pH adjustments have to be made for natural water samples. The powder is insensitive to substances like alkali and alkaline earth ions and humic matter. Improvements in the method of preparation of the 2,2′-diaminodiethylaminecellulose material led to a reduction of the metal blank and consequently to a reduction in element detection limits to the 0.1 μg $litre^{-1}$ level, compared to 0.5 μg $litre^{-1}$ for the earlier modified cellulose disc procedure. In the revised method[59] the sample (200 ml) is filtered through a Nucleopore 0.4 μm pore size membrane, in order to remove all suspended and precipitated material, 50 mg 2,2′-diaminodiethylaminecellulose powder is then filtered off again on a Nucleopore 0.4 μm filter in a Nuclepore filtration unit with 12 cm active diameter.

Figure 3 of Volume 1 shows the effect of sample pH on the collection efficiency of the cellulose powder for several elements. The results imply that, except for manganese(II), all the cations under examination can be preconcentrated effectively at the natural pH of environmental water, i.e. without the previous addition of any pH adjusting reagents.

Hyphan cellulose (Cellex P) and pyrogallol cellulose. Burba *et al.*[60] used these celluloses to preconcentrate, by factors of 100–200, nanogram quantities of beryllium in natural water including sea-water prior to determination by graphite furnace atomic absorption spectrometry.

Storage of natural water samples:

Natural water samples (saline and non-saline): Collect in precleaned polyethylene bottles. After separation of suspended matter (0.45 μm filtration) acidify to a pH of less than 2 with concentrated suprapure nitric acid (2 ml nitric acid per litre sample).

Reagents/equipment:

Chelating cellulose ion exchangers: Cellulose Hyphan (TM of the Riedel de Haen AG, Hannover/Seelze, FRGO).[61] For purification, treat the ion exchangers successively with 4 M hydrochloric acid, tridistilled water, and reagent grade acetone. The air-dried ion exchangers in the H^+ form were stable for months.

Hydrochloric acid spec. pure.

Graphite furnace AAS. Perkin Elmer HGA 74, HGA 76 B and HGA 500. The optimized temperature program is shown in Table 8.

Table 8(a) Atomic absorption spectrometer operating conditions.

Wavelength	234.9
Slit	0.5 nm
Source	Be hallow-cathode lamp
Lamp current	5 mA
Burner (FAAS)	One-slit burner for the acetylene–nitrous oxide flame
Flame	C_2H_2/N_2O, stoichiometric
Background	D_2-background correction.

(b) Graphite-furnace program for the Be determination

Program step	Temperature/Time
Drying	100 °C/30 s
Ashing	1000 °C/20 s
Atomization[a]	2700 °C/10 s (mini flow)
Clean-up	2700 °C/15 s

[a]With D_2-background correction.
From Burba *et al.*[60] with permission.

Organically bound beryllium:

To release beryllium bound to organic matter irradiate polluted water samples with a UV lamp[63] (Hg low pressure lamp 70 W, 1 h at 45 °C in a quartz reactor).

Procedure:

Batch method: To 200 ml of the stored water sample add 0.1 g cellulose pyrogallol. Then, under magnetic stirring adjust the pH of the solution to 8.5 with 2 M sodium hydroxide solution add 5 ml borate buffer solution, after 45 min filter the loaded ion exchanger on a filter paper, wash, and dry at 90°C. Elute beryllium with 2 ml 2 M hydrochloric acid.

Column method: Suspend 0.3 g Hyphan then 0.3 g pyrogallol cellulose (capacity 0.4 mmol g^{-1}) in high pure water. Load into a glass column with an inner diameter of 1 cm. The ion exchanger had a volume of about 2.5 ml in the wet state resulting in a filling height of approximately 3 cm. Adjust the pH of the solution to 8.5 by adding 5 to 10 ml pH 8.5 borate buffer. Use a flow rate of 5 ml min^{-1}. After separation wash with 10 ml high purity water. Elute beryllium with 5 ml 2 M hydrochloric acid. Run reagent blanks through the whole analytical procedure, and deduct from the beryllium content found in the sample.

Natural levels of beryllium found by this method ranged from less than 1 μg $litre^{-1}$ in sea-water with a 90 per cent recovery to–14μg $litre^{-1}$ in river water, with a 95 per cent recovery. Less than 1 μg $litre^{-1}$ beryllium was found in potable water samples. Both hyphan cellulose and pyrogallol cellulose were equally effective in preconcentrating beryllium.

Cellulose hyphan has been used for the preconcentration of lead, copper, nickel, cadmium, and zinc from water.[64] The pH of the samples was shown not to be critical over the range 5–8. The metal ions could be eluted from the column with 1 M nitric acid. No interference was caused by salts commonly present in natural potable waters. Burba *et al.*[65] evaluated this resin for the determination of trace uranium in natural and potable waters using atomic absorption spectrophotometry, inductively coupled plasma emission spectrometry, and X-ray fluorescence spectromentry for the analytical finish. Down to μg quantities of uranium can be determined by this procedure, without interference from inorganic matter or organic matter including humic acid and lignin. Following the enrichment, traces of uranium are directly determined on the ion excharger by wavelength dispersive X-ray fluorescence analysis (detection limit (3α) 1.5 μg $litre^{-1}$ U). On the other hand, after elution by 2 M hydrochloric acid uranium can be determined by means of spectrophotometry (U-complex of Arsenazo(III) or Chlorphosphonazo(III), detection limit 0.1 μg $litre^{-1}$ U, at 10 μg $litre^{-1}$ U) or ICPAES (detection limit 0.5 μg $litre^{-1}$ U, at 10 μg $litre^{-1}$ U).

Natural cellulose. Whilst this material is not suitable for the preconcentration of metals from water it does have some sorbtive properties for metals occurring in the enviroment and these have been studied.[69] Distribution coefficients of various metal ions on different natural celluloses are tabulated in Table 9. The distribution coefficients of about 10^5 that can be reached on cellulose in alkaline solutions (pH 10–12) are very interesting for analytical trace enrichment. Another remarkable result is the very different sorption of chromium(III) and chromium(V) species on cellulose. For example, it makes possible their analytical separation and speciation in natural waters.

The K_d values found for individual species, e.g. copper and zinc, at neutral pH (e.g. pH 8) in 0.5 M sodium chloride prove to be nearly constant (e.g. 10^3 for the two just mentioned) over large concentration ranges (e.g. from 200 ng to 1 mg $litre^{-1}$). However, in the presence of other trace metals (e.g. iron(III) at concentrations of some 100 μg $litre^{-1}$) a considerable synergistic enhancement of the distribution coefficients occurs. Figure 5 of Volume 1 shows the influence of electrolyte concentration (sodium chloride, 0–30 g $litre^{-1}$) on the heavy metal sorbtion (Cd^{2+}, Co^{2+}, Fe^{3+} Ni^{2+}, Zn^{2+} each at the 20μl level) on cellulose at pH 11. In the absence of sodium chloride the distribution

Table 9 Distribution coefficients (w/v) on different celluloses for metal ions at the 20 μg litre^{-1} level (pH 11.0).

Element	log K_d					
	Cellulose S & S[a] 123/3	Cellulose S & S[a] 123ag	Linters S & S[a] 124a	Powder S & S[a] 180a	Powder Avicel[κ] Merck 2330	Native cellulose Merck 2351
Cd^{2+}	3.73	3.58	3.51	3.80	3.85	3.74
Co^{2+}	4.99	4.30	3.85	4.60	4.57	4.98
Cu^{2+}	4.25	4.12	4.10	4.50	3.75	4.75
Fe^{3+}	≥5.0	≥5.0	≥5.0	≥5.0	≥5.0	≥5.0
Mn^{2+}	4.0	3.95	3.72	4.05	3.94	4.40
Pb^{2+}	3.67	3.90	3.70	3.80	3.85	3.85

[a] Schleicher & Schüll.
From Burba *et al.*[65] with permission.

coefficients reach values of 500–2000 with the exception of Fe^{3+}. However, even a 1.5–2 g litre^{-1} concentration of sodium chloride will enhance the K_d values by a factor of about 30. At higher salt levels the distribution equilibria seem to be nearly constant. Similar salt effects can be observed in neutral suspensions of cellulose. In contrast to these results, other materials investigated as sorbents in natural waters tend to give remobilization of fixed heavy metal traces at higher salt concentrations.

p-Phenylendiaminecellulose. *p*-Phenylenediaminecellulose has been used for the preconcentration of mercury in water.[67] *o*-Aminophenolcellulose ether and glyoxal-bis(2-mercaptoaniline)cellulose both chelate mercury(I), mercury(II), methylmercury, and cadmium(II)[68] which can be subsequently desorbed.

Uron and induron loaded cellulose. Extremely low concentrations of molybdenum and vanadium in water samples can be preconcentrated by coprecipitation of the elements from acid solution into uron or indiuron loaded celluloses.[69]

1.1.4 Chemically modified polymers—sea-water

Imido diacetate resins, Chelex-100 Pai *et al.*[70] examined the preconcentration efficiency of Chelex-100 for heavy metals in sea-water. They showed that the metal-chelating efficiency of this resin was lower in sea-water than in fresh water owing to the complicated speciation of metals in sea-water and competition from high levels of magnesium and calcium. The optimal pH for sea-water samples prior to loading was about 6.5, but optimal pH values for column operation were strongly affected by the salt matrix. Care should be exercised in the choice of conditions to minimize losses of cadmium and manganese from sea-water. Pai[71] also examined the distribution of heavy metals in sea-water on a Chelex-100 column. Columns were divided into small sections and spiked heavy metals recovered from each section. A simplified model was derived to describe the effect of column size and flow rate on the recovery of heavy metals (chelating efficiency) at various pH values. Columns containing Chelex 100 resin (2 g) in the magnesium form with a flow rate of 4 ml min^{-1} were suitable for the preconcentration of cadmium, copper, cobalt, manganese, nickel, lead, and zinc in sea-water after adjustment of samples to pH 6.5.

Atomic absorption spectrometry. Studies of the use of ion exchange resins for the preconcentrations of metals from sea-water have been mainly concerned with the use of Chelex-100 cation exchange resin.[74–76] Chelex-100 is the most commonly employed chelating resin fow the removal and preconcentration of trace heavy metals from sea-water. Work on the use of Chelex-100 cation exchange resin for the preconcentration of metals from sea-water is reviewed in Table 10. In each case metals are desorbed from the resin with nitric acid (2–2.5 M) and then determined in the extract by graphite furnace atomic absorption spectrometry. Preconcentration factors of up to 100–120[72–74] have been reported by this technique enabling metals to be determined at the μg $litre^{-1}$ of ng $litre^{-1}$ level.

Early Chelex-100 procedures only partially separated the alkali and alkaline earth metal components of sea-water prior to the analysis of the eluted elements of interest. A more recent separation procedure utilizing the Chelex resin produced a sample devoid of alkali, alkaline earth, and halogen elements, and left a dilute nitric acid/ammonium nitrate matrix containing only the trace elements of the sea-water samples.[72] While this procedure produces a highly desirable and appropriate matrix for most spectroscopic methods of analysis, a solid sample would be more appropriate for other instrumental techniques such as X-ray fluorescence or neutron activation analysis. In addition, the above separation procedure also makes it difficult or impossible to analyse several elements which are held strongly by the resin but cannot be quantitatively eluted. Chromium and vanadium exhibit this type of behaviour and attempts to reproducibly elute these elements from Chelex-100 have not been successful.

Paulson[77] studied the effects of flow rate and pretreatment on the extraction of manganese, cadmium, and copper from estuarine and coastal sea-water by Chelex-100 resin. Decreasing the flow rate for column extraction of estuarine samples by Chelex-100 to 0.2 ml min^{-1}

Table 10 Application of Chelex-100 resin to the preconcentration of metals in sea-water prior to analysis by graphite furnace atomic spectrometry.

Elements	Concentration factor	Eluent	Detection limit	Ref.
Cd, Co, Cu, Fe, Mn, Ni, Pb, Zn	100:1	2.5 M nitric acid	subanogram $\mu g\ litre^{-1}$	72
Cu, Cd, Zn, Ni	120:1	2 M nitric acid	Cu 0.006	73
			Cd 0.006	
			Zn 0.015	
			Ni 0.015	
Cd, Zn, Pb, Cu, Fe, Mn, Co, Cr, Ni	20:1	2.5 M nitric acid	Not stated $\mu g\ litre^{-1}$	47
Cd, Pb, Ni, Cu, Zn	100:1	2.5 M nitric acid	Cd 0.01	75
			Pb 0.16–0.28	
			Ni 0.24–0.68	
			Cu 0.6	
			Zn 1.8	

increases the yield of trace metals and improves the precision for determination of these elements. Detection limits achievable by chelation–solvent extraction and ion exchange separation on Chelex-100 resin are summarized in Table 11.

Kingston *et al.*[78] and Bender and co-workers[79,80] used Chelex-100 followed by stripping with nitric acid to preconcentrate total and soluble manganese from sea-water. Samples were collected into 500-ml polyethylene bottles. All samples were brought to pH 2 with nitric acid free of trace metals and stored in individual zip-lock plastic bags to minimize contamination. When the samples were returned to the laboratory the pH was adjusted to approximately pH 8 using concentrated ammonia (ultrapure, G. Frederick Smith). Chelex-100 cation exchange resin in the ammonia form (20 ml) was added to the samples and they were batch extracted on a shaker table for 36 h. The resin was decanted into columns and the manganese eluted using 2 N nitric acid.[78] The eluant was then analysed by graphite furnace atomic absorption spectrophotometry. Replicate analyses of samples indicate a precision of about 5 per cent.

Batley and Farrah[81] and Gardner and Yates[82] used ozone to decompose organic matter in samples and thus break down metal complexes prior to atomic absorption spectrometry. By this treatment metal complexes of humic acid and EDTA were broken down in less than 2 min. These observations led Gardner and Yates[82] to propose the following method for the determination of cadmium in sea-water: The sample is filtered immediately after collection, acidified to about pH 2, and transferred to a 1-litre Pyrex storage bottle. Prior to extraction the sample is ozonized in the sample bottle for 30 min. Nitrogen is passed through the sample for 5 min to remove excess ozone, then the pH is carefully raised to about 5 by addition of ammonia solution and about 5 ml Chelex-100 resin in the ammonia form is added. After stirring for at least 1 h, the resin is collected in a Pyrex chromatography column and washed with the calculated quantity of an appropriate buffer to elute calcium and magnesium. After further washing with 50 ml deionized water, the resin is eluted with 2 M nitric acid, to a volume of 25 ml. The eluate is analysed by graphite furnace atomic absorption spectrometry.

Boniforti *et al.*[83] compared several preconcentration methods in the determination of metals in sea-water. A comparison was made of ammonium pyrrolidinedithiocarbamate–8-quinolinol complexation followed by extraction with methyl isobutyl ketone or Freon-113, coprecipitation with magnesium hydroxide or iron(II) hydroxide or chelating by batch treatment with Chelex-100 for the determination of manganese, iron, cobalt, zinc, nickel, copper, and chromium. Atomic absorption spectrometry and inductively coupled plasma atomic emission spectrometry were used for analysis. Interferences, recovery, preci-

Table 11 Applicability of preconcentration procedures to sea-water analysis. Detection limits (based on 2 SD of blank).

Element	Concentration in sea-water concensus values (μg litre^{-1})	Best achievable values by preconcentration		
		Chelation solvent extraction–graphite furnace AAS	Ion-exchange separation Chelax-100 resins	
			Graphite furnace AAS	Neutron activation analysis
Cr	0.03	0.05	–	0.14
Mn	0.02	0.004	–	0.16
Fe	0.2	0.02	–	1.2
Co	0.005	0.04	–	0.006
Ni	0.17	0.012	0.015	–
Cu	0.05	0.006	0.006	0.08
Zn	0.49	0.016	0.015	0.20
Cd	–	0.0001	0.0006	–
Pb	–	0.016	–	–
Th	0.01			0.0004
V	2.5			0.06

From Paulson[77] with permission.

sion, accuracy, and detection limits were compared. The Chelex-100 resin method was most suitable for the preconcentration of all determinands except chromium, whereas preconcentration of Cr^{III} and Cr^{V} was achieved only by coprecipitation with iron(II) hydroxide.

Olsen *et al.*[84] gave details of equipment and procedure developed for preconcentrating and determining traces of cadmium, lead, copper, and zinc in sea-water by atomic absorption spectrophotometry combined with flow injection analysis. Preconcentration was achieved by passing 2 ml of sample through a microcolumn packed with Chelex-100 resin. Metals were desorbed with 180 μl 2 M aqueous nitric acid which was passed to an atomic absorption spectrometer, thereby achieving a concentration factor of about 10. Seen from a practical viewpoint this combination results in timesaving because it allows an unprecedented sample throughput at the μg litre^{-1} level. As the analytical readout is available within 5 s for the direct assay and the latest within 110 s for the system including preconcentration, smaller sample series can be treated expediently by manual injection.

Since the original work of Olsen many improvements have been proposed to flow injection analysis equipment. Valve designs have been improved, in addition to Chelex-100 other ion exchange resins have been tested, and different flow systems have been proposed to increase the efficiency of the process.[85-88] These developments indicate that the new approach not only increases the speed of the preconcentration process, but could also ultimately rival the sensitivity and speed of graphite furnace atomic absorption spectrometry.

Fang *et al.*[89] have described a flow-injection system with on-line ion exchange preconcentration on dual columns for the determination of trace amounts of heavy metals at μg litre^{-1} and sub-μg litre^{-1} levels by flame atomic absorption spectrometry. The degree of preconcentration ranges from 50- to 105-fold for different elements at a sampling frequency of 60 samples hourly. The detection limits for copper, zinc, lead, and cadmium are 0.07, 0.03, 0.5 and 0.05 μg litre^{-1}, respectively. Relative standard deviations were 1.2–3.2 per cent at μg litre^{-1} levels. These workers studied the behaviour of the different chelating exchangers used with respect to their preconcentration characteristics, with special emphasis on interferences encountered in the analysis of sea-water.

Fang studied closely the swelling properties of Chelex-100 resin, and showed that depending on the pH of the element, the resin swelled by a factor up to two. To limit the maximum change in volume of a resin in a column to 25 per cent during a single cycle of operation (which is imperative to avoid excessive pressure and void volume variations), the sampling period (during which chelation takes place) at a flow rate of 6 ml min^{-1} should not exceed 50 s for a 0.5 M ammonium acetate buffer, 75 s for a 0.1 M buffer or 100 s for a 0.05 M buffer. Secondly, the column should be packed about three-quarters full with resin in the H^+ form (washed with water) after conversion from the NH_4^+ form (in which state it should be equilibrated with buffer of the same concentration as that used in the ensuing procedure). Packing in the NH_4^+ form would otherwise result in an excessively loose column packing giving rise to large dispersion and degrading the sensitivity. Finally, resin columns in the NH_4^+ form should never be washed with water, lest excessive pressure develop in the column, causing blockages, leakages or dislodgement of the nylon retaining gauzes. When not in use, the columns should be converted to the H^+ form by normal elution with acid and washed thoroughly with water.

Fang concluded that the flow-injection atomic absorption spectrometry system with on-line preconcentration will challenge the position of the graphite furnace technique, because it yields comparable sensitivity for much lower cost by using simpler apparatus and separation mode. The method offers unusual advantages when matrices with high salt contents such as sea-water are analysed because the matrix components do not reach the nebulizer.

Inductively coupled plasma atomic emission spectrometry. Berman *et al.*[90] attempted to determine nine elements in sea-water by a combination of ion exchange preconcentration on Chelex-100[91,92] and ICPAES using ultrasonic nebulization. Preconcentration factors of between 25 and 100 were obtained by this technique. Table 12 compares the ICPAES results with data generated for the same sample by two other independent methods—isotope dilution spark source mass spectrometry (IDSSMS) and graphite furnace atomic absorption spectrometry

Table 12 Analysis of Sandy Cove sea-water.

Element	ICPAES	GFAAS	IDSSMS
Mn	1.5 ± 0.1	1.1 ± 0.2	1.8 ± 0.2[a]
Fe	1.5 ± 0.6	1.5 ± 0.1	1.4 ± 0.1
Zn	1.5 ± 0.4	1.9 ± 0.2	1.6 ± 0.1
Cu	0.7 ± 0.2	0.6 ± 0.2	0.7 ± 0.1
Ni	0.4 ± 0.1	0.33 ± 0.08	0.37 ± 0.02
Pb	–	0.22 ± 0.04	0.35 ± 0.03
Cd	–	0.24 ± 0.04	0.28 ± 0.02
Cr	–	–	0.34 ± 0.06
Co	–	–	0.02[a]

Precision expressed as 95 per cent confidence intervals.
[a]Spark source spectrometry, internal standard method.
From Berman *et al.*[90] with permission

(GFAAS). The IDSSMS method also uses a preconcentration of the metals and matrix separation using the ion exchange procedure, following isotope addition. The atomic absorption determinations were preceded by an MIBK extraction.[91] In general the agreement is very good, but one discrepancy merits comment. The spark source mass spectrometry result for manganese is not as reliable as the other data by this method. Since manganese is monoisotopic, a less accurate internal standardization method of calibration has to be used. The ICPAES result for manganese is in close agreement with the GFAAS result.

Manganese, iron, zinc, copper, and nickel can be determined by preconcentration ICPAES giving good agreement with the other methods. No consistently reliable results were obtained for cadmium, chromium, cobalt, and lead when samples up to 1 litre were processed (i.e. no preconcentration). Chromium is only weakly retained by the Chelex-100 resin under the conditions used so that the sea-water concentration (about 0.3 μg litre^{-1}) is not sufficiently enhanced.

Neutron activation analysis. The application of neutron activation analysis to water samples has several advantages. Like inductively coupled plasma atomic absorption spectrometry it is a multi-element analysis technique and, as such, is useful for performing element scans when looking for unknown elements. It enables several elements to be determined in the same run and is not limited to metallic elements. Its disadvantage is high cost and the probability that most laboratories cannot house the reactor equipment so that, in fact, samples have to be sent away for analysis to an establishment with such facilities. This delays results and demands careful consideration of sample preservation during the waiting period. The technique is intrinsically sensitive, and

Table 13 Trace elements in water—SRM 1643a (μg litre^{-1}).

Element	Neutron activation analysis[a]	Certified[b]
Co	19.1 ± 1	19 ± 2
Cr	16 ± 2	17 ± 2
Cu	19.1 ± 0.6	18 ± 2
Fe	88 ± 16	88 ± 4
Mn	30.9 ± 0.6	31 ± 2
Mo	97 ± 6	95 ± 6
Ni	56 ± 8	55 ± 3
V	52 ± 1	53 ± 3
Zn	68 ± 5	72 ± 4

[a]Uncertainties are 2 SD.
[b]Uncertainties are 95 per cent confidence limit.
From Greenberg and Kingston[93] with permission.

can be made more so by preconcentrating the metals on to a resin such as Chelex-100 which can be directly analysed by the neutron activation technique. With these facilities analysis for many metals at the ultra-low background levels at which they occur in sea-water becomes a possibility.

Greenberg and Kingston[93] described a method to prepare solid samples from 100 ml estuarine or sea-water using Chelex-100 resin, followed by the determination of trace elements in the solid resin by neutron activation analysis. Using this procedure, typical decontamination factors of $\geqslant 10^7$ for sodium, $\geqslant 10^5$ for chlorine, and $\geqslant 10^3$ for bromine are observed. They used this procedure to determine cobalt, chromium, copper, iron, manganese, molybdenum, nickel, scandium, thorium, uranium, vanadium, and zinc in NBS. Standard Reference Material 1643a (Trace Elements in Water) (Table 13). They also analysed a sample of sea-water taken in Chesapeake Bay by neutron activation analysis and compared these results with those obtained by other techniques. The good agreement is evident in Table 14.

In a procedure[94] employing preconcentration of the metals on a column containing a mixture of Chelex-100 and Pyrex glass powder the problems associated with swelling of the Chelex-100 were overcome and constant flow rates of sample down the column achieved. The water sample were passed through the resin column and eluted with 100 ml 0.01 M nitric acid and the eluate was discarded. Trace elements were collected from the column by eluting with 50 ml 4 M nitric acid. The eluate was heated to reduce its volume, transferred to a volumetric flask, and diluted to 10 ml.

Table 14 Trace elements in sea-water sample (ng ml^{-1})[a].

Element	Neutron activation analysis	GFAAS	XRF
Co	0.044 ± 0.003	<0.1	
Cr	3.3 ± 0.2		
Cu	2.01 ± 0.05	2.0 ± 0.1	2.0 ± 0.2
Fe	2.1 ± 0.3	2.1 ± 0.5	
Mn	1.89 ± 0.03	2.0 ± 0.1	2.0 ± 0.1
Mo	5.3 ± 0.1		
Ni	1.3 ± 0.2	1.2 ± 0.1	1.3 ± 0.2
Sc	<0.00095 ± 0.0005		
Th	<0.0002		
U	1.90 ± 0.04		
V	0.45 ± 0.01		
Zn	4.9 ± 0.2	4.8 ± 0.3	4.5 ± 0.4

[a] All uncertainties are at the ISD level.
GFAAS = Graphite furnace atomic absorption spectrometry. XRF = X-ray fluorescence spectrography.
From Greenberg and Kingston[93] with permission.

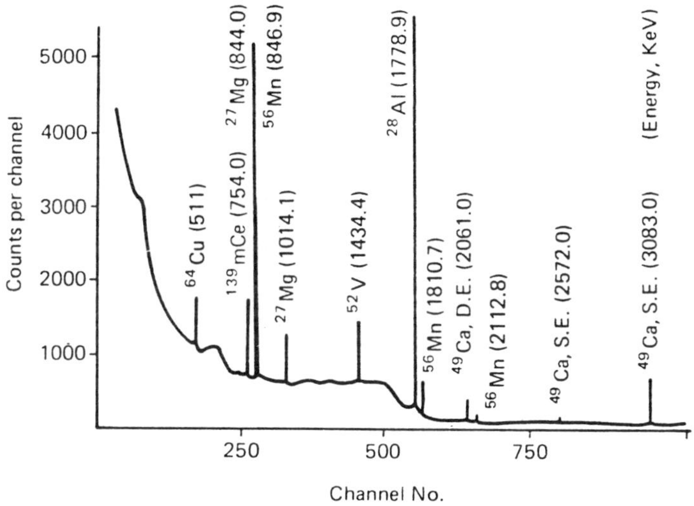

Fig. 1 Gamma-ray spectrum of preconcentrated river water after short irradiation. Irradiation time 3 min; thermal-neutron flux 1×10^{13} n cm^{-2}, s^{-1} decay time 3 min; counting time 100 s. From Lee *et al.*[94] with permission.

A 3.5 ml portion in a 4 ml polyethylene vial was irradiated for 5 min (Fig. 1). Another portion, 3.0 ml in a 3.5 ml silica vial, was irradiated for 3 days. After the short irradiation, 3 ml of the irradiated solution were transferred into an activity-free vial and submitted to γ-ray spectrometry with a Ge(Li) detector coupled with the 4000 channel analyser. After the long irradiation the sample was allowed to cool for 3 days, then the surface of the silica ampoule was cleaned with dilute nitric acid, and the sealed ampoule was placed in the counter (the background activity of the ampoule was negligible). γ-Ray energy and peak-areas were calculated by computer. To determine the half-lives of the nuclides produced, the counting was repeated at appropriate intervals.

The recovery ratios shown in Tables 15 and 16 indicate that the added traces of cadmium, cerium, copper, lanthanum, manganese, scandium, and zinc are quantitatively recovered. The recoveries of barium, cobalt, bromium, iron, uranium, and vanadium were also satisfactory. Table 16 shows the recoveries of calcium and magnesium were very poor for sea-water. The reasons for this may be connected with matrix effects. In order to evaluate the precision of this method, replicate analyses were carried out by Lee *et al.*[94] by the proposed procedure, for trace elements in a sea-water sample taken from the Kwangyang Bay. The results in Table 17 show satisfactory precision.

Table 15 Separation of interfering elements from trace elements per cent recovery.

Isotopes	From effluent (2 litres)	From successive 20 ml portions of 0.01 M HNO_3					From successive 10 ml portions of 4 M HNO_3					Total
		1	2	3	4	5	1	2	3	4	5	
^{24}Na	95.1	3.9	0.5	0.3	0.1	0	0	0	0	0	0	99.9
^{82}Br	98.2	1.3	0.4	0.1	0	0	0	0	0	0	0	100.0
^{42}K	96.3	3.0	0.5	0.4	0.1	0	0	0	0	0	0	100.3
^{38}Cl	97.0	2.4	0.6	0.2	0	0	0	0	0	0	0	100.2
^{115m}Cd	0	0	0	0	0	0	18.1	37.0	38.4	6.2	0.3	100.0
^{64}Cu	0	0	0	0	0	1.4	31.2	43.1	21.7	2.5	0	99.9
^{56}Mn	0	0	0	0	0	0	21.8	33.0	32.7	10.3	1.9	99.7
^{64}Zn	0	0	0	0	0	0	19.9	38.1	39.9	2.2	0.1	100.2

From Lee *et al.*[94] with permission.

Table 16 Recovery of trace elements from spiked samples of stripped waters.

Element	Sea-water		
	Added (μg)	Found (μg)	Yield (per cent)
Ba	1.44	1.48	103
Ca	–	–	–
Cd	30.0	27.6	90.0
Ce	1.48	1.45	98.0
Co	1.50	1.49	99.3
Cr	230	230	100
Cu	189	187	98.9
Fe	615	615	100
La	0.830	0.832	100
Mg	–	–	–
Sb	–	–	–
Sc	0.140	0.142	101
U	0.160	0.157	98.1
V	27.0	27.0	100
Zn	401	397	99.0

From Lee *et al.*[94] with permission.

Table 17 Results of replicate analysis for trace elements in waters taken from the Kwangyang Bay and the Nakdong River, South Korea.

Element	Sea-water (μg litre^{-1})
Ba	4.8 ± 0.8
Ca	–
Cd	0.20 ± 0.05
Ce	16.7 ± 0.7
Co	–
Cr	2.33 ± 0.20
Cu	1.1 ± 0.07
Fe	250 ± 10
La	0.72 ± 0.04
Mg	–
Mn	1.50 ± 0.04
Sc	0.098 ± 0.004
U	1.36 ± 0.1
V	2.14 ± 0.05
Zn	45.9 ± 1.4

From Lee *et al.*[94] with permission.

Table 18 Application of neutron activation analysis to the determination of metals in sea-water.

Elements	Sample preconcentration	Ref
Co, Cr, Ca, Fe, Pb, Se, Sr	Sea salts	99
	Freeze dried	
As, Cu, Sb	Thionalide precipitation	100
Misc elements	Frozen sample	101
Al, V, Cu, Mo, Zn, U	Organic coprecipitation (8-hydroxyquinoline)	102
Ag, Au, Cd, Ce, Co, Cr, Eu, Fe, Hg, La, Mo, Sc, Se, U, Zn, As, Sb	Adsorption and charcoal	103
Ba, Ca, Cd, Ce, Co, Cr, Cu, Fe, La Mg, Mn, Se, U, V, Zn	Adsorption of Chelex-100 glass powder	104
Hg, Au, Cu	Coprecipitation with lead diethyldithiocarbamate	105
12 elements	Chelating resin	106
Transition metals	Chelating resin	107
As, Mo, U, V	Colloid flotation	108
Th	Ion exchange	109
Co, Cu, Hg	Coprecipitation with lead pyrrolidine dithiocarbamate	110
In	Ion exchange chromatography	111
Co	Ion exchange	112

Some further applications of neutron activation analysis to sea-water analysis are summarized in Table 18.

Stable isotope dilution spark source mass spectrometry. Mykytiuk *et al.*[109] have described a stable isotope dilution spark source mass spectrometric method for the determination of cadmium, zinc, copper, nickel, lead, uranium, and iron in sea-water and compared results with those obtained by graphite furnace atomic absorption spectrometry and inductively coupled plasma emission spectrometry. These workers found that to achieve the required sensitivity it was necessary to preconcentrate elements in the sea-water using Chelex-100[110] followed by evaporation of the desorbed metal concentrate on to a graphite or silver electrode for isotope dilution more spectrometry.

Results obtained on a sea-water sample by three procedures are given in Table 19. Isotope dilution results (IDSSMS) agree well with those obtained by graphite furnace atomic absorption spectrometry (GFAAS) and inductively coupled plasma emission spectrometry (ICPES). The preconcentration for all three techniques was achieved by Chelex-100 ion exchange. However, since solvent extraction with ammonium pyrrolidinedithiocarbamate is the most commonly accepted method, the values obtained using it with graphite furnace atomic absorption spectrometry are also included for comparison.

One of the advantages of the isotope dilution technique is that the quantitative recovery of the analytes is not required. Since it is only their isotope ratios that are being measured, it is necessary only to recover sufficient analyte to make an adequate measurement. Therefore, when this technique is used in conjunction with graphite furnace

Table 19 Analysis of sea-water sample B (concentrations (ng ml^{-1}) expressed as means $\pm$SD).

	IDSSMS ion exchange	GFAAS		ICPES ion exchange
		Solvent extraction	Ion exchange	
Fe	3.4 ± 0.3	3.2 ± 0.2	3.4 ± 0.4	3.2 ± 0.2
Cd	0.07 ± 0.01	0.06 ± 0.01	0.053 ± 0.007	ND
Zn	1.9 ± 0.1	1.8 ± 0.1	2.0 ± 0.1	1.6 ± 0.2
Cu	0.61 ± 0.04	0.5 ± 0.1	0.51 ± 0.03	0.73 ± 0.06
Ni	0.43 ± 0.03	0.46 ± 0.03	0.45 ± 0.05	0.38 ± 0.02
Pb	0.11 ± 0.02	0.06 ± 0.02	0.10 ± 0.01	ND
Co	0.028 ± 0.001[a]	0.015 ± 0.003	0.018 ± 0.008	ND
U	2.6 ± 0.2	ND	ND	ND

[a] By internal standard.
ND = not determined.
From Mykytiuk *et al.*[109] with permission.

atomic absorption spectrometry, it is possible to determine the efficiency of the preconcentration step. This is particularly important in the analysis of sea-water where the recovery is very difficult to determine by other techniques since the concentration of the unrecovered analyte is so low. In using this technique, one must assume that isotopic equilibrium has been achieved with the analyte regardless of the species in which it may exist.

Flame photometry. Blake *et al.*[111] have described a flame photometric method for the preconcentration of calcium in solutions of high sodium content. The method was applied to simulated sea-water. In the method Chelex-100 chelating resin (Na^+ form) was stirred with 2 N hydrochloric acid for 5 min, the acid was decanted and the resin washed with water, stirred with 2 N sodium hydroxide for 5 min, and again washed with water. The procedure was repeated five times then the resin was dried at 100 °C. A neutral solution containing up to 50 μg litre^{-1} of calcium and up to 4 per cent of sodium was passed through a column of the resin, a specified amount of hydrochloric acid (pH 2.4) was passed through and the percolate containing the sodium was discarded. Elution was then effected with 2 N hydrochloric acid (5 ml) and the column washed with water (25 ml), the combined eluate and washings were diluted to 100 ml and calcium determined by flame photometry at 622 nm. There was no interference from magnesium, zinc, nickel, barium, mercury, manganese, copper or iron present separately in concentrations of 25 μg litre^{-1} or collectively in concentrations of 5 μg litre^{-1} each. Aluminium depresses the amount of calcium found.

Dithiocarbamate substituted celluloses A limited amount of work has been carried out on the determination of molybdenum in sea-water by atomic absorption spectrometry[112-114] and graphite furnace atomic absorption spectrometry.[115] In a recommended procedure[115] a 50 ml sample of sea-water at pH 2.5 is preconcentrated on a column of 0.5 g *p*-aminobenzylcellulose, then the column is left in contact with 1 M ammonium carbonate for 3 h, after which three 5 ml fractions are collected. Finally, molybdenum is determined by atomic absorption at 313.2 nm with use of the hot-graphite-rod-technique. At the 10 mg litre^{-1} level the standard deviation was 0.13 μg.

Murthy and Ryan[116] preconcentrated copper, cadmium, mercury, and lead from sea-water on a column of a dithiocarbamate cellulose derivative. Metal concentrations on the adsorbent material were determined by neutron activation analysis. The recovery of added spikes to sea and tap water suggests that Cu^{II}, Hg^{II}, and Pb^{II} can be quantitatively collected.

Dithiocarbamate substituted Chromosorb Knockel and Prange[117] converted metals in sea-water into their diethyldithiocarbamates prior to

X-ray fluorescence analysis of the preconcentrated solids. Membrane filtration of the precipitates resulted in carbamate-loaded filters, which could be directly measured by using radioisotope-excited X-ray fluorescence analysis. Furthermore, elution of Chromosorb columns loaded with the dithiocarbamate complex, by the passage of chloroform gave chloroform solutions in which the trace metals could be determined by X-ray fluorescence analysis using totally reflecting sample supports. Similarly the precipitate on the membrane filter could be dissolved in chloroform and determined in solution. The sensitivity of this method and the pH dependence of the reaction was also investigated.

1.1.5 Polymeric adsorbents

Polyurethane foam

Non-saline waters. Omar and Bower[118] have investigated a procedure for the preconcentration and determination of tin in lake and natural waters which involves preliminary preconcentration by adsorption of tin onto polyurethane foam soaked in toluene-3,4-dithiol. A column of polyurethane foam extracts tin(II) from natural waters over the pH range 2–8. The tin is converted to the tetravalent state to remove interferences and determined spectrophotometrically at 650 nm as its complex with catechol violet and cetyltrimethyl ammonium bromide.

Apparatus:

Unican SP600 spectrophotometer or equivalent.

pH meter.

Filtration equipment. Millipore filters (0.45 μm, 5 cm in diameter) supported by a sintered glass base (Millipore, Bedford, MA, USA)

Reagents:

Tin(IV) stock solution (0.2 mg ml^{-1}). Dissolve 50.0 mg of pure tin in 20 ml of hot concentrated sulphuric acid. Cool, add 100 ml of water and 30 ml of concentrated sulphuric acid, then make up the volume to 250 ml.

Polyurethane foam–dithiol. Cubes of polyurethane foam (Vert Foam components, Manchester) were soaked in a 1 per cent m/v solution of toluene −3,4-dithiol in diethyl ether and squeezed dry between watch glasses before use.

Catechol violet 0.02 per cent m/v in water.

Cetyltrimethylammonium bromide, 0.01 m/v in water.

Potassium iodide, a 5 M solution freshly prepared.

Washing solution, 5 ml of 5 M potassium iodide solution added to 25 ml of 4.5 M sulphuric acid.

Procedure:

Preconcentration of inorganic tin from natural waters: Adjust a 1 litre sample of water to pH 4 and treat with 1 ml of mercaptoacetic acid which is sufficient to reduce any tin(IV) to tin(II). Pass the sample through a glass column, 15 cm long and 1.5 cm in diameter, containing 0.4 g of polyurethane foam–dithiol, at a rate of 3–5 ml min^{-1}. This size of column will retain up to 15 μg of tin quantitatively. (Other elements found to be retained to some extent were antimony(I), cobalt, iron(III), silver, and zinc. The retention of these is reduced to less than 0.5 per cent of the input by adding 0.1 g of disodium ethylenediaminetetracetate to the sample.)

After passing the water sample through the column, which takes 3–4 h, recover the tin by eluting the column with a mixture of acetone and 1 M sodium hydroxide solution, (3 + 2 v/v) at a rate of 2 ml min^{-1}. The first 2.5 ml of eluate contains no tin and is rejected. The next 10 ml contains all of the tin free from organic sulphides, which can interfere with the catechol violet determination described below. In this way, inorganic tin is concentrated 100-fold with a net loss of 2 ± 1 per cent but is not freed from all interfering substances.

Separation of tin from interfering substances by solvent extraction: Heat 10 ml of eluate from the previous step under an infra-red lamp to volatilize the acetone and then wet ash. Dissolve the residue in 20 ml of 5 M sulphuric acid and transfer into a separating funnel. Samples not requiring preconcentration need to be dissolved in 20 ml of 5 M sulphuric acid and the tin oxidized to tin(IV). Then add 5 ml of 5 M potassium iodide solution and shake the mixture with 10 ml of toluene for 1 min. Discard the aqueous layer and wash the toluene layer with 10 ml of washing solution without shaking. Discard the washings, and transfer the toluene layer into a clean separating funnel. Shake it with 5 ml of 1 M hydrochloric acid, collect the aqueous layer and wash the toluene with 2 ml of 0.1 M hydrochloric acid, add the washings to the aqueous phase and discard the toluene.

Determination of tin by spectrophotometry: To the sample in 7 ml of approximately 1 M hydrochloric acid, add 0.3 ml of ascorbic acid, 1.5 ml of catechol violet, and 1 ml of acetyltrimethylammonium bromide shaking the solution after each addition. Add about 10 ml of water and adjust the pH to 2.4 ± 0.1 with 2 M sodium hydroxide solution. Make the solution up to 25 ml with water and allow it to stand for 15 min. Measure the absorbance at 650 nm in a 1 cm cell using water as a reference. Wash out the cell with dilute hydrochloric acid followed by distilled water and correct the absorbance for any mismatched cells and reagent blanks. The calibration graph of absorbance against concentration is linear over the range 0.04–4 μg of tin.

The extraction of metals from water samples by polyurethane foam has been discussed by several other workers.[119-121] Cobalt was preconcentrated by percolating the water sample through an open cell polyurethane support loaded with 1-(2-pyridylazo)naphthol prior to spectrophotometric determination at 510 nm with the 4-(2-pyridylazo)resorcinol complex. Various heavy metals have been precipitated[121] on to polyurethane discs of foam loaded with ammonium diethyldithiocarbamate to increase the sensitivity of XRF spectrometry and the foam subsequently analysed by X-ray fluorescence spectrometry. Gawargious *et al.*[122] adsorbed the coloured bismith iodide complex from potable water on to thin-lavers of polyurethane foam prior to measurement of its absorbance at 495 and 600 nm against a blank foam in benzene. Recoveries were between 103 and 107 per cent. Copper and silver interfered.

Polyethylene

Non-saline waters. Testa and Staccioli[123] have pointed out that Microlthene-710 (a microporous polyethylene) as a support material for triphenylphosphine oxide in cyclohexane medium has a potential application for the determination of plutonium in fall-out samples.

Sea-water. Delle Site *et al.*[124] have used extraction chromatography to determine plutonium in sea-water, sediments, and marine organisms. These workers used double extraction chromatography with Microthenes-210 (microporous polyethylene) supporting tri-*n*-octylphosphine oxide (TOPO); a technique that has been used previously to isolate plutonium from other biological and environmental samples.[125] Plutonium-236 and plutonium-242 were tested as the internal standards to determine the overall plutonium recovery, but plutonium-242 was generally preferred because plutonium-236 has a shorter half-life and an α-emission (5.77 MeV) which interferes strongly with the 5.68 MeV (95 per cent) α-line of ^{224}Ra, the daughter of ^{228}Th. However, the 5.42-MeV α-lines of ^{228}Th interfere with those of ^{238}Pu (5.50 MeV) and so a complete purification from thorium isotopes is required.

Plutonium sources were counted by an α-spectrometer with good resolution, background, and counting yield. The counting apparatus used had a resolution of 40 KeV. The mean ($\pm$so) background value was 0.0004 ± 0.0003 cpm in the ^{239,240}Pu energy range and 0.0001 ± 0.0001 cpm in the ^{238}Pu energy range. The mean ($\pm$so) counting yield, obtained with ^{239}Pu, ^{240}Pu reference sources counted in the same geometry, was found to be 25.08 ± 0.72 per cent.

To determine the overall recovery obtained by this procedure (chemical recovery and electrodeposition yield) a known activity of ^{242}Pu was added to the different samples; the plutonium sources were counted by α-spectrometry for 3000 min and the percentage overall recovery was calculated from the area of ^{242}Pu peak. The percentage overall recovery for 3-litre samples of sea-water was 63 ± 10 per cent. Owing to the very low activity of the samples, the determination of ^{239}Pu, ^{240}Pu, and ^{238}Pu in the reagents is very important in calculating the net activity of the radionuclides.

The method proposed was checked by analysing some sea-water reference samples prepared by the IAEA Marine Radioactivity Laboratory (Monaco) for intercomparison programmes. The values reported by IAEA and the experimental values obtained by Delle Site[124] were in good agreement.

1.1.6 Polymeric resins—non-saline waters

Carboxylated polyethylenimine–polymethylenephenylene isocyanate The resin was used for preconcentrating metals from estuary and sea waters[126] prior to analysis by inductively coupled plasma spectrometry. The uptake of copper, cadmium, lead, and zinc by the resin was quantitative in the presence of acetate and citrate buffers. The collection of other heavy metals and rare earths was also investigated.

Polyamine polyurea resin Trace heavy metals (copper, nickel, zinc, and cobalt) can be preconcentrated[127] from natural waters on to this resin. Using the tetraethylenepentamine resin coated on to glass beads heavy metals can be concentrated from 10–20 litres of aqueous sample with 75 to 95 per cent efficiency achieving concentration factors of up to 1000. The final analysis was carried out by atomic absorption spectrometry.

Aniline S resin Fitzgerald *et al.*[128] reported a cold trap preconcentration technique for the determination of trace amounts of mercury in water. Kramer and Neidhart[129] determined μg $litre^{-1}$ levels of mercury by using an aniline sulphur resin for the selective enrichment of mercury from surface waters.

Polystyrene supported poly(maleic anhydride) resin This resin has been used to preconcentrate lead in potable water prior to determination by flame atomic absorption spectrometry.[130] Chelex-100 in the calcium form proved unsuitable for lead preconcentration. This method met the European Community directive limiting the content of lead in potable water to 50 μg $litre^{-1}$.[131]

A plumbo solvent potable water having a low alkalinity (18 mg $litre^{-1}$ as $CaCO_3$) and low pH (6.58) had most of the lead in the smaller size fractions but a significant proportion is particulate (i.e. >0.4 μm). The Ca–Chelex-100 extraction was not 100 per cent efficient for lead removal from any of the subsamples. Incomplete recoveries may be due to the presence of colloidal species or to non-Chelex labile organic complexes of lead or both. Slow dissociation of complexes relative to the solution/resin contact time may also have an effect. The polystyrene supported poly(maleic anhydride) resin behaved in a significantly different manner. The column extraction (50 ml of sample) gave 100 per cent lead removal from the 0.4 and 0.08 μm filtered waters but particulate lead was not recovered from the unfiltered and 12 μm filtered samples. A batch extraction on the unfiltered sample after passage through the column chelated all the remaining lead from solution thereby indicating its potential for total lead determinations.

Some potable water samples of relatively high alkalinity were examined because such waters can attack lead. Total lead in unfiltered samples was determined by AAS directly and following a 48 h batch extraction of 50 ml of water. Lead removal by the poly(maleicanhydride) resin was complete for both high and low alkalinity samples.

Macroreticular polyacrylic acid and SM-7 ester resin Corsini *et al.*[132] have used a column of this resin to preconcentrate unchelated trace metals in water samples prior to determination by atomic absorption spectrometry. It is claimed to be simple, rapid, and economical. A study of the effect of pH on the uptake of cadmium(II), cobalt(II), copper(II), manganese(II), nickel(II), lead (II), chromium(III), and iron(II) showed that with 100 ml of sample containing 1 ml 1 M ammonium acetate buffer for the divalent ions at levels ranging from 1–20 μg litre^{-1} in water a maximum metal uptake (85–100 per cent) occurred in the pH range 7–9, and for the trivalent ions in the pH range 4–5. Nitric acid 1 per cent was the most satisfactory eluting agent for removing the metals from the column. More concentrated solutions of nitric acid (e.g. 10 per cent) increase the blank value considerably due to contamination. For a 7 cm bed height, the chromatographic elution profiles of all ions are essentially the same.

Analytical results for the lake water sample are given in Table 20. Data obtained for preconcentration by SM-7 and Chelex-100 are compared. Included are data obtained by direct analysis. The concentration of some elements present is high enough for measurement by graphite furnace atomic absorption without preconcentration. The method of standard additions was used for all data in Table 20. The results represent total soluble metal. The agreement among the methods is acceptable in consideration of the low concentration involved.

On the whole, the blank values for the SM-7 results were 2–4 times lower than with Chelex-100, the blank for iron(III) in particular being most improved. The lower blanks arise because the SM-7 resin is cleaner and because less nitric acid is required for elution. The lesser amount

Table 20 Comparison of SM-7 and Chelex-100 methods in analysis of lake-water (results in μg litre^{-1}).

	SM-7[a]	Chelex-100	Direct
Fe^{III}	2.4 ± 0.4	2.3	2.4
Cr^{III}	0.9 ± 0.1	0.79	1.0
Pb^{II}	0.12 ± 0.01	0.14	<0.2
Cu^{II}	1.4 ± 0.1	1.1	1.5
Ni^{II}	2.0 ± 0.2	2.3	2.6
Co^{II}	0.09 ± 0.01	0.05	<0.2
Mn^{II}	0.60 ± 0.01	0.39	0.4
Cd^{II}	0.072 ± 0.001	0.093	<0.1

[a] SM-7 data are the average of two results; data for Chelex-100 and direct analysis represent a single determination. Flow rates for uptake and elution for both columns were 1.0–1.2 ml min^{-1}.
From Corsini *et al.*[132] with permission.

of nitric acid may be at least in part a consequence of a lower affinity of the SM-7 resin than that of the Chelex-100 for metal ions, in view of the fact that metal ion uptake by a strong chelating mechanism cannot occur with SM-7. The disadvantages is that two sample aliquots are used, one for uptake of trivalent ions at pH 4–5 and another for divalent ions at pH 8–9.

Whereas percentage recovery of metal ion spikes from distilled water were essentially quantitative by the above method, spike recoveries from lake water were considerably lower than quantitative for example 80 per cent recovery of copper to as low as 48 per cent recovery of lead. Subsequently these workers[133] devised a two column procedure for preconcentrating these metals from lake water and sea-water in which organic ligands, particularly humic materials, were removed by passing the water through a precolumn at pH 1–2 before preconcentrating the trace metals at pH 8. The final effluent for measurement by graphite furnace atomic absorption spectrometry was readily matrix matched and allowed the use of the standard calibration curve procedure for the majority of the trace metals.

1.1.7 Membrane filters

Taguchi *et al.*[134] preconcentrated cadmium on a nitrocellulose membrane filter as the 1-(2-pyridylazo)-2-naphthol complex. This chelate was then dissolved in 0.5 ml of warm concentrated sulphuric acid together with the membrane filter. No interferences were observed in the presence of high levels of sodium chloride, calcium, magnesium, sulphate, and bromide or in the presence of sub-mg litre^{-1} levels of zinc(II), iron(III), manganese(II), and copper(II). The lower limit of detection for cadmium was 1.5 ng litre^{-1}.

Ohzeki *et al.*[135] preconcentrated trace amounts of aluminium from potable water by collection on a nitrocellulose membrane filter, prior to spectrophotometric determination using Chrome Azurol S and Zephiramine. Zephiramine, Chrome Azurol S and sodium perchlorate were added to an acidified solution of aluminium(III). The aluminium(III)–Chrome Azurol S–Zephiramine complex and the Zephiramine–perchlorate ion-pair precipitate were collected on a membrane filter by filtration under suction. The selectivity of the method was increased by adding a ligand buffer solution of *trans*-cyclohexane-1,2-diaminetetra-acetic acid. The solid-state absorbance of the complex was measured at 640 and 700 nm against the blank thin layer and the difference calculated. The method could be used to determine levels of filterable aluminium in tap water (about 2.3 ng ml^{-1}) with reproducible results.

1.1.8 Bentonite

Mundschenk[136] enriched strontium-90 and yttrium-90 in natural waters by precipitation as the phosphates and adsorption on to a bentonite matrix. This enabled down to 0.02 pCi litre^{-1} of these elements to be estimated using 100-litre samples.

1.2 Anions – non-saline waters

1.2.1 Sephadex

Yoshimura *et al.*[137] have described a highly sensitive method for preconcentrating borate based on adsorption on Sephadex G-25 gel in alkaline medium and reversible desorption in acid medium. The borate is then determined spectrophotometrically by the azomethine-4 method.

1.2.2 Nitrocellulose and acetylcellulose membranes

A preconcentration technique based on these materials has been described for phosphate.[138] Phosphorus was collected as phosphomolybdate on a nitrocellulose or acetylcellulose membrane in the presence of *n*-dodecyltrimethylammonium bromide. The membrane was dissolved in dimethyl sulphoxide and the absorbance of dimethyl sulphoxide solution measured. Moderate concentrations of silicate, anionic and non-ionic surfactants and high concentrations of sodium chloride did not interfere. Arsenate interference could be eliminated by reducing arsenate to arsenite. Determination of condensed and organic phosphates was possible following their conversion to orthophosphoric acid. The limit of determination was 0.02 μg litre^{-1} phosphate.

1.2.3 2,2′-diethylaminocellulose filters

2,2′-diethylaminocellulose filters enabled selenate (SeONH$_4^{2-}$) and selenite (SeO$_3^{2-}$) to be preconcentrated from natural water at pH 3–6 with a detection limit of 0.05 μg litre^{-1}.[139,140]

1.2.4 Polyurethane foam

Farag *et al.*[141] determined chromium(VI) (chromate) in water using 1,5-diphenylcarbazide loaded polyurethane foam.

Polyurethane foam has also been used to preconcentrate phosphate. Khan and Chow[142] have described a method in which the phosphate is converted into phosphomolybdate which is then extracted into polyurethane foam and analysed for molybdenum by X-ray fluorescence spectrometry directly on the foam. A polyether type polyurethane foam disc is squeezed for an hour in a mixture of phosphate, sodium molybdate, and hydrochloric acid, spiked with phosphorus-32. After washing and drying, the foam disc is placed on plastic foam and stretched across the X-ray source holder. The method is simple and rapid and the precision is 5 per cent for 0.25 mg litre^{-1} and 2 per cent for 2.5 mg litre^{-1} phosphate. Equimolar amounts of silicon, germanium, and arsenic(IV) appear to interfere with the determination.

Palagni[143] used the pulsed column bed technique to selectively separate and preconcentrate low levels of iodide-127 in natural water. The pulsed bed consisted of a syringe containing a scintillation detector and a gamma ray spectrometer. The adsorbent consisted of open cell polyurethane foam impregnated with long chain tri-*n*-alkylamine which formed a complex with iodide.

1.3 Organics

1.3.1 Polyurethane foam

Hydrocarbons Ahmed *et al.*[153] investigated the precision and accuracy of infra-red spectroscopic methods for determining oil in sea-water using two techniques of oil preconcentration, namely carbon tetrachloride extraction and polyurethane foam adsorption. The foam adsorption method of Schatzberg and Jackson[154] was used. Estuary

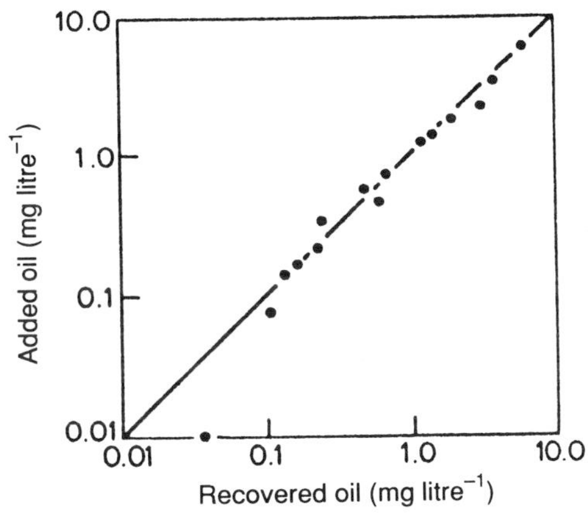

Fig. 2 Oil concentration as determined by infra-red spectrometric technique versus known concentration. From Ahmed *et al.*[153] with permission.

water (10–15) litres was passed through a foam disc (8 cm diameter, 4 cm thick, 100 pores in^{-1}) and then the retained oil was extracted in a Soxhlet apparatus with carbon tetrachloride. This extract was measured as described above. It was necessary to clean new discs by Soxhlet extraction with carbon tetrachloride for 4–6 h to reduce blank analysis to acceptable levels. Recoveries of known concentrations of oil were greater than 85 per cent for those concentrations above 5 mg $litre^{-1}$.

The results in Fig. 2 obtained on synthetic solutions of oil in carbon tetrachloride show that the average recovery is 101 per cent and the average error is 13 per cent. Therefore, the method itself seems quite reliable since the quantity of oil in the sample bottle can be determined with a relative standard deviation of less than 15 per cent for concentrations exceeding about 0.05 mg $litre^{-1}$. Replication experiments on samples gave relative standard deviations for each set of replicates from 46 to 130 per cent with the average being 75 per cent; this variability is far in excess of the variability of the analytical procedure itself (15 per cent). This error must, therefore, be associated with sampling from a heterogeneous system such as oil in water.

Comparative date for the extraction and foam adsorption preconcentration methods are given in Table 21. In all cases, the oil concentration resulting from the foam techniques is the lower of the two values; in fact, it is lower by a factor of five on the average. Although there are sampling errors associated with both techniques it is clear that the foam technique has a much lower collection efficiency than liquid–liquid extraction and is, therefore, a less useful procedure.

Table 21 Comparison of oil concentrations determined using different preconcentration techniques (surface sample only).

CCl_4 extraction (mg litre^{-1})	Foam adsorption (mg litre^{-1})	Ratio
84.1	7.2	11.7
0.10	0.08	1.25
0.21	0.03	7.0
0.283	0.19	1.5
0.239	0.07	3.4
		Av. 5.0

From Ahmed *et al.*[153] with permission.

Carsin[155] has carried out a literature review of the preconcentration of hydrocarbons from polluted water. He reviewed sampling equipment used in the determination of hydrocarbon contaminants in natural waters. A new type of sampler in which a floating collector is employed, and an aspirator pump draws the contaminated water through a bed of oil-absorbent polyurethane foam. Problems of operation of the sampling apparatus, including the behaviour at sea under different weather conditions are discussed. The recovery and subsequent analysis of the oil fraction by infra-red spectrometry are also described.

Polychlorinated biphenyls Porous polyurethane foam has also been used to preconcentrate polychlorobiphenyls from water samples. Gesser *et al.*[156] found that the compounds could be adsorbed on a column composed of two polyurethane plugs (each 38 mm × 22 mm) inside a glass tube. The sample water was poured through at 250 ml min^{-1} then the plugs were removed and squeezed free from water; the PCBs were then extracted by treating the plugs with acetone and hexane. The concentrated extract was analysed by gas chromatography on a glass column (6 ft × 0.25 in) packed with Chromosorb W HMDS supporting 2 per cent SE-30 and 3 per cent of QF-1 and operated at 200 °C; the carrier gas was helium and a ^{63}Ni electron capture detector was used. In tests on 20 μg of polychlorobiphenyls added to 1 litre of water, the recovery (based on measurements of 13 peaks on the chromatogram) ranged from 18.2 to 19.6 μg. Bedford[157] also used polyurethane foam plugs to extract polychlorobiphenyls from natural waters; these results indicated that Aroclor is probably adsorbed on to small particles in unfiltered lake water which can pass through the foam. Although this makes the foam method somewhat impractical for enriched and turbid waters, it is still a valuable technique for clear water containing low quantities of polychlorobiphenyls.

Polyurethane foam has been used as a liquid–liquid partitioning filter for the concentration of polychlorobiphenyls from water samples.[158,159]

Polyaromatic hydrocarbons Saxena *et al.*[160] used polyurethane foams to preconcentrate trace quantities of six representatives of polynuclear aromatic hydrocarbons (fluoranthene, benzo(k)fluoranthene, benzo(j)fluoranthene, benzo(a)pyrene, benzo(ghi)perylene, and indeno (1,2,3-cd)pyrene) prior to regular screening of these compounds in US raw and potable waters. Final purification and resolution of samples was by two-dimensional thin layer chromatography followed by fluorimetric analysis and quantification.

In this method the polyaromatic hydrocarbons are collected by passing water through polyurethane foam plugs. Water is heated to 62 ± 2 °C prior to passage and flow rate is maintained at approximately 250 ml min^{-1} to obtain quantitative recoveries. The collection is followed by elution of foam plugs with organic solvent, purification by partitioning with solvents and column chromatography on Florisil. Analysis is by two-dimensional thin layer chromatography on cellulose acetate–alumina plates followed by fluorimetry and gas liquid chromatography using flame ionization detection. The latter method was less sensitive than thin layer chromatography. Employing this method and a sample volume of 60 litres, polyaromatic hydrocarbons were detected in all the water supplies sampled. Although the sum of the six representative polyaromatic hydrocarbons in potable waters was small (0.9–15 μg $litre^{-1}$) the values found for raw waters were as high as 600 μg $litre^{-1}$.

In further work Saxena *et al.*[161] and Basu and Saxena[162] showed that the polyurethane foam plug method had an extraction efficiency for polyaromatic hydrocarbons of at least 88 per cent from treated

Table 22 Foam retention efficiencies for PAH from potable water.

Compound	Amount added to water (μg)	Per cent retention
FL	100	100
BjF	100	88^{b}
BkF	100	
BaP	100	81
IP	100	89
BghiP	100	91

^{a}FL, fluoranthrene; BjF, benzo(j) fluoranthrene; BkF, benzo(k) fluoranthrene; BaP, benzo(a) pyrene; IP, indeno(1,2,3cd) pyrene; BghiP, benzo(ghi) pyrene.
^{b}Combined value given since the compounds could not be separated on the GLC column.
Water source, laboratory tap water; water volume, 4 litre; conc. of each PAH, 25 ppb; detection method, GLC–FID.
From Saxena *et al.*[161] with permission.

waters and 72 per cent from raw waters. Foam retention efficiencies of the six polyaromatic hydrocarbons from spiked laboratory potable water at 25 μg litre^{-1} are shown in Table 22. The data confirm that polyurethane foam plugs under suitable conditions not only effectively concentrate benzo(a)pyrene but other polyaromatic hydrocarbons as well. Foam plugs concentrated polyaromatic hydrocarbon almost quantitatively from finished water at lower concentrations also (Table 23). The high polyaromatic hydrocarbon retention efficiencies were also maintained with heavily polluted surface waters (Table 24). It is unclear why the retention values for three of the polyaromatic hydrocarbons are well above 100 per cent. A possible explanation for this may be the inability of cyclohexane extraction to quantitatively recover these

Table 23 Foam retention efficiencies for PAH from treated water.

Compound	Conc. in aqueous phase (ng litre^{-1})	Amount retained by foam from 1 litre of water (ng)	Per cent retention
FL	278.6	260.4	93.5
BjF	48.3	47.4	98.1
BkF	51.7	50.6	97.6
BaP	36.4	33.6	92.3
IP	25.5	23.9	93.7
BghiP	22.6	19.8	87.6

[a] See Table 22.
Water source, laboratory tap water; water volume, 60 litres; conc. of fluoranthene, 500 ppt; all others, 100 ppt; detection method, TLC–fluorimetric.
From Saxena *et al.*[161] with permission.

Table 24 Foam retention efficiencies for PAH from raw water.

Compound	Conc. in aqueous phase (ng litre^{-1})	Amount retained by foam from 1 litre of water (ng)	Per cent retention
FL	289.1	343.7	118.9
BjF	77.6	94.0	121.1
BkF	66.1	55.6	84.1
BaP	74.5	59.7	80.1
IP	85.2	61.2	71.8
BghiP	23.9	28.3	118.4

[a] See Table 22.
Water source, Onondaga Lake water; volume, 30 litres; Conc. of fluoranthene, 500 ppt; all others, 100 ppt; detection method, TLF–fluorimetric.
From Saxena *et al.*[161] with permission.

Table 25 Limit of detection of PAH with foam preconcentration coupled with ILC–fluorimetric or GLC–FID method.

Compound[a]	TLC–fluorimetric detection		GLC–FID detection	
	Absolute limit (ng)	Limit in 60 litres water (ng $litre^{-1}$)	Absolute limit (ng)	Limit in 60 litres water (ng $litre^{-1}$)
Fl	140.0	2.3	13.6	4.5
BjF	7.5	0.1	10.1	3.4
BkF	5.0	0.1		
BaP	10.0	0.2	11.9	4.0
IP	10.0	0.2	14.7	4.9
BghiP	20.0	0.3	14.9	5.0

[a]See Table 22.
From Saxena *et al.*[161] with permission.

polyaromatic hydrocarbons from the heavily polluted water, this will give rise to lower polyaromatic hydrocarbon concentration in the aqueous phase than actually present.

Table 25 shows the detection limits of the six polyaromatic hydrocarbons using the polyurethane foam preconcentration method obtained by gas chromatography with a flame ionization detector or thin layer chromatography fluorimetric analysis. The detection of fluoranthene and benzo(a)pyrene in thin layer chromatography fluorimetry is restricted by the background levels of these compounds contributed from the foam plugs. Their detection limits are assumed to be twice the background fluorescence level. In the case of the gas chromatographic FID method, the detection limits for polyaromatic hydrocarbons are based on a minimum output response of five times the background noise level and a maximum volume of 5 μl from a total of 100 μl concentrate. Details of this procedure are given below.

During the concentration of polyaromatic hydrocarbons from water on foam plugs, several other contaminants were also concentrated and some of these were eluted during polyaromatic hydrocarbon elution. In addition several impurities originating in the foam were also leached during the elution process. The latter impurities could be partially eliminated only be precleaning of the plugs with cyclohexane and/or benzene by batch or Soxhlet extraction. The impurities interfered with the analysis of polyaromatic hydrocarbons, therefore a clean-up procedure was devised. The levels of impurities and subsequently the extent of the clean-up necessary are directly dependent upon the sample volume and the number of foam plugs employed for concentration. A sample volume of 60 litres was adequate for detection of polyaromatic hydrocarbon. Sixty litres of unspiked finished water was passed through six precleaned foam columns each containing two plugs, in three successive steps maintaining the water temperature at 62 ± 2 °C and flow rate at 250 ± 10 ml min^{-1}. Each column was eluted with 30 ml acetone and 125 ml cyclohexane. The combined extract was concentrated and subjected to clean-up. At no time was the polyaromatic hydrocarbon mixture allowed to proceed to complete dryness since this has been shown to result in loss of polyaromatic hydrocarbons.[170]

Additional clean-up involving a short Florisil column was found to be necessary for further separation from impurities. Chromatographic grade Florisil (60–100 mesh) was washed with methanol and 1:1 hexane-benzene and activated for at least 4 h at 130 °C. The Florisil was cooled to room temperature and 8 g was transferred to a glass column (1.5 × 30 cm) with benzene by slurry method and washed with an addition of 100 ml of benzene prior to passing sample concentrate through it. The flow chart illustrated in Fig. 3 depicts the complete clean up procedure. Thin layer chromatography was performed using aluminium oxide acetylated cellulose plates, 2 per cent (w/v) of $CaSO_4 \cdot 2H_2O$ (200 mesh) was added to the slurry to increase binding of the layer to the surface. The plates were developed in two dimensions, *n*-hexane–benzene

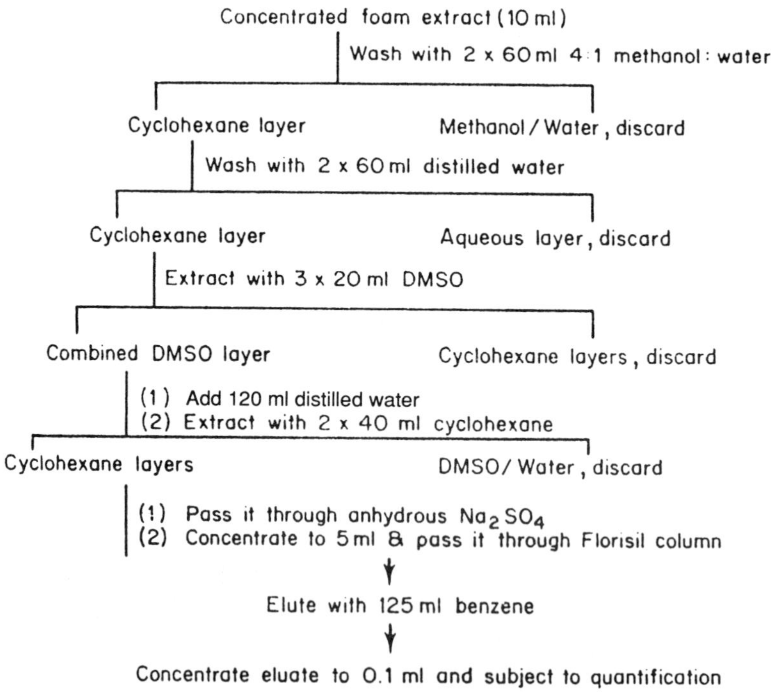

Fig. 3 Flow chart of clean up method. From Saxena *et al.*[161] with permission.

(4:1 v/v) and methanol–ether–water (4:4:1 v/v).

The emission and excitation spectra used for identification of the suspected polyaromatic hydrocarbon spots were run directly on the plates at room temperature with a thin film scanner attached to a spectrophotofluorimeter. This procedure eliminated the losses usually encountered during removal of polyaromatic hydrocarbon spots for fluorescence measurement in solvents and the interferences arising due to solvent interaction. For quantification of the spots, the excitation wavelength was fixed at 365 nm and the fluorescence intensities were measured at the following wavelengths (nm): fluoranthene (458), benzo(j)fluoranthrene (427), benzo(k)fluoranthene (428), benzo(a)pyrene (427), and benzo(ghi)pyrene (416).

Navra'til *et al.*[164] have also investigated the use of high capacity open pore polyurethane columns for the collection and preconcentration of polyaromatic hydrocarbons from water. They tested open pore polyurethane columns for ion exchange properties, solvent compatabilities, and ability to remove and concentrate polynuclear aromatic hydrocarbons from water. Their performance was compared with that of other sorbent materials using aqueous solutions of pyrene.

Afghan *et al.*[165] present results from studies on the feasibility of using polyurethane foams to extract polynuclear aromatic hydrocarbons from water prior to their determination. They also studied factors affecting adsorption. Preliminary filtration of the sample to remove particulate matter is recommended. Polyurethane foam is particularly useful when large samples are routinely analysed in that it offers high flow rates, good capacities, and low costs compared to solvent extraction. Polyaromatic hydrocarbons can be extracted at the part per trillion (pg ml^{-1}) from natural waters.

1.3.2 Polyethylene and polypropylene

Non-saline waters

Chlorinated insecticides. Thin polyethylene film has been employed for the preconcentration of traces of chlorinated insecticides from water. Weil[166,167] used 20–25 μm thick film, and presents results for adsorption of γ-BHC, heptachlor expoxide, methoxychlor, dieldrin, and DDT. Experiments[168] with ^{14}C-lindane and ^{14}C-DDT showed that adsorption of these substances on polyethylene film was not likely to be of practical use in river water analyses because of the effects of foreign matter. Low surface area polypropylene has been evaluated[169] as an adsorbent for preconcentrating aqueous organic compounds. An extraction process was developed and the adsorbent capacity of polypropylene was tested with 19 compounds (polycyclic aromatic hydrocarbons, ketones, esters, alcohols, alkanes). Results indicated an inverse relation between solubility and retention on the adsorbent. The surface micro environment of the polymer was investigated by using pyrene as a molecular fluorescence probe.

1.3.3 PTFE (Chromosorb T)

Non-saline waters Josefson *et al.*[170] investigated the adsorption of trace organic compounds from water by Chromosorb T, an aggregate of polytetrafluoroethylene, in column chromatography. The differences in adsorption capacity between Chromosorb T and another polytetrafluoroethylene aggregate, Fluoropak 80, was attributed to differences in surface morphology. The adsorption of 33 organic solutes in aqueous solution was studied by frontal chromatography and the capacity was shown to be inversely related to solubility of solute for polycyclic aromatic hydrocarbons with the exception of the xanthines. The use of Chromosorb T columns to recover hydrophobic solutes from synthetic hard water containing such solutes was found to be effective.

1.3.4 Support bonded silicones

Non-saline waters Aue *et al.*[171] used support bonded silicones for the preconcentration of 10–20 ng litre^{-1} of lindane, heptachlor, aldrin, heptachlor epoxide, and dieldrin prior to gas chromatography. This absorbent was packed on to a glass tube (35 × 1 cm) and 10 litres of sample (or of pure water treated with known compounds) was passed through the column at 50–55 ml min^{-1}; the column was then dried by passage of nitrogen. The sorbed compounds were eluted with pentane (2 × 5 ml) and a portion of the elute was injected directly into a borosilicate glass column (170 cm × 3.5 mm) packed with 1.5 per cent of QF-1 plus 2 per cent of OV-17 on Chromosorb W-HP (100–120 mesh). The column was operated at 185 °C for chlorinated insecticides with nitrogen (60 ml min^{-1}) as carrier gas and a ^{63}Ni electron capture detector.

1.3.5 Sephadex

Non-saline waters Adsorption rates of various organic compounds on this gel have been studied by Tombo *et al.*[144] (G-15 gel), Hiraide *et al.*[145] (Sephadex A25), and Hine and Bursill.[146] Hiraide *et al.*[145] studied complexes of humic and fulvic acid with divalent copper, lead, and cadmium in pond water. The negatively charged humic substances were selectively and quantitatively sorbed from 1 litre samples on a small column of the macroreticular, weak-base, anion exchanger, diethylaminoethyl-Sephadex A-25. The complexed metals were then quantitatively desorbed with a small volume of 4 M nitric acid and determined by graphite furnace atomic absorption spectrometry. The method was simple, rapid, and gave reproducible and reliable results. About 80 per cent of the humic substances sorbed on the Sephadex column could be eluted with 0.5 M sodium hydroxide and determined spectrophotometrically at 400 nm.

Hezzlar[147] studied the effect of inorganic salts on the adsorption of organic substances on Sephadex gel from samples of peat-bog water, river water, and secondary sewage effluent. Most of the organic compounds possessed pH-dependent reversible adsorption affinity to the gel. This effect was attributed to the presence of inorganic salts; the possible mechanism involved is discussed.

Sea-water Minear[148] and Minear and Walonski[149] used Sephadex gel to preconcentrate naturally occurring soluble organophosphorus compounds from sea-water.

1.3.6 Vinylpyridine–divinylbenzene copolymer

Non-saline waters Kawabata and Ohira[172] have studied the preconcentration of phenols from aquatic samples using vinylpyridine–divinylbenzene copolymer as an adsorbent. Although the analytical implications of this work were not discussed these clearly exist. Elution of the concentrated phenols from the resin column was accomplished by a treatment with acetone or methanol.

1.3.7 N-Vinyl-2-pyrrolidone copolymer

Non-saline waters Carpenter *et al.*[173] preconcentrated phenolic materials from aqueous solutions in cross linked aqueous insoluble *N*-vinyl-2-pyrrolidone polymer. The pH for maximal binding of the phenolic compound to the resin was found to be dependent on the acidity of the phenolic compound. Binding to resin was particularly favourable for polyhydroxyl and extended aromatic compounds. Columns packed with this resin moved more than 95 per cent of simple phenolic compounds from aqueous solution and quantitative recovery of the bound phenolic compound was possible by elution with 4 M urea solution.

1.3.8 Octadecyl resin

Non-saline waters O'Keefe *et al.*[174] have described a sampling cartridge containing octadecyl reverse phase adsorbent, for collecting of 2,3,7,8-tetrachlorodibenzo-*p*-dioxin from potable water. Procedure for subsequent clean-up and analysis of the samples is also described and results of tests to assess the efficiency of the method are included. The lower limit of detection for 2,3,7,8-tetrachlorodibenzo-*p*-dioxin by this method was about 1 pg $litre^{-1}$.

1.3.9 Polymethacrylic ester

Non-saline waters Stuberu and Leenheer[175] assessed the aqueous elution, selective concentration approach for isolating aromatic bases from water, and the factors controlling the concentration process identified. The degree of concentration attainable depends on the ratio of the capacity of the neutral form of the amine to that of the ionized form. The capacity factors of ionic forms of amines on polymethylacrylic ester resins are 20–250 times lower than those of the neutral forms and increase with the hydrophobicity of the amine.

1.3.10 Acrylonitrile–divinylbenzene copolymer

Non-saline waters Acrylonitrile–divinylbenzene copolymer has been used to preconcentrate methanol, acetaldehyde, hexafluoroacetone, and acetic acid from natural water[176] prior to determination by gas chromatography.

1.3.11 Styrene–divinylbenzene copolymer and Sep-pak C-18

Non-saline waters Robinson *et al.*[177] evaluated disposable cartridges containing styrene–divinylbenzene copolymer hydrophobic resin (Chrom-Prep RRP-1) (containing octadecyl bonded phase silica) for preconcentration of selected aromatic hydrocarbons, phthalate esters, and food, drug, and cosmetic (FD and C) dyes in aqueous solutions. Chrom-Prep cartridges were compatible with eluents and samples over the pH range 1–13 whereas Sep-Pak cartridges were limited to pH 1–7. The former had a greater absorption capacity for model aromatic hydrocarbons and phthalate eaters, whereas the latter exhibited greater capacity for the FD and C dyes, which had breakthrough volumes that were a function of pH. Analysis of aqueous samples spiked with ^{14}C-labelled solute standards showed that both cartridge types leached a small amount of analyte during the trace enrichment step.

1.3.12 1,4-di(methacryloyloxymethyl)naphthalene–divinylbenzene copolymer

Non-saline waters Gawdzik and Gawdzik[178] preconcentrated chlorophenols by solid phase extraction using this copolymer. They compared recoveries and breakthrough volumes of phenol and several mono-, di-, and tri-chlorophenols for this copolymer and several other chemically bonded phases containing hexyl and octadecyl groups (alkylsilane modified silica RP-6 and RP-18). 1,4-di(methacryloyloxymethyl)naphthalene–divinylbenzene copolymer gave yields of about 100 per cent for aqueous solutions containing 2 μg ml^{-1} of each study compound. This level of recovery was achieved for RP-18 only with 2,4,6-trichlorophenol and not at all for RP-6.

1.3.13 Lipidex 5000

Non-saline waters Pentachlorophenol has been preconcentrated from natural water on a Lipidex 5000 bed.[179] Pentachlorophenol was eluted with acetone and simultaneously separated from lipophilic compounds. It was analysed by GC with electron capture detection after derivatization to pentachlorophenyl acetate. Less polar compounds could also be determined. For example, 1,1-bis(4-chlorophenyl)-2,2,-trichloroethane (*p,p'*-DDT) and 2,3,7,8-tetrachlorodibenzo-*p*-dioxin (TCDD) were collected in a separate fraction from the gel. Average recovery of 0.1–0.15 ng pentachlorophenol per ml water was 96 per cent and that of labelled *p,p'*-DDT at 5 ng ml^{-1} and TCDD at 0.001 ng ml^{-1} were 95 and 92 per cent, respectively.

1.3.14 Wofatit Y77 polymeric adsorbent

Non-saline waters Dedek *et al.*[180] used the new polymeric sorbent Wofatit Y77 to preconcentrate hydrophilic insecticides and herbicides from water or saturated sodium chloride solution. Radiolabelled pesticides were then desorbed using methanol and the radioactivity measured in appropriate aliquots of the eluate. Recoveries were in excess of 90 per cent for phosphorus-32 labelled methamidophos and carbon-14 labelled demethoate, trichlorfon, sodium trichloroacetate, fenuron, propachlor, and 2,4-D. Hot solvents were recommended for the elution of the more strongly adsorbed hydrophobic compounds such as 2,4-D.

1.3.15 Chromosorb W

Non-saline waters Chromosorb W has been used to preconcentrate polychlorinated biphenyls prior to gas chromatographic analysis.[181] Ahling and Jensen[150] used a column of Carbomex 4000 monostearate/Chromosorb W to selectively preconcentrate chlorinated insecticides from samples also containing polychlorinated biphenyls. The insecticides were then desorbed with a small volume of organic solvent prior to analysis by gas chromatography.

1.3.16 Cellulose acetate

Non-saline waters Kurtz[151] and Musty and Nickless[152] have used cellulose triacetate to adsorb polychlorinated biphenyls.

1.4 Organometallic compounds

1.4.1 Polyurethane foam

Non-saline waters

Mercury. Phenylmercury and methylmercury compounds have been preconcentrated on diethylammonium diethyldithiocarbamate loaded polyurethane foam prior to determination by X-ray fluorescence analysis.[181] Braun *et al.*[182] showed that polyurethane foam loaded with diethyldithiocarbamate is suitable for the preconcentration of trace amounts of organic and inorganic mercury prior to its chemical analysis.

1.4.2 Sulphydryl cotton fibre

Non-saline waters Lee[184] preconcentrated methyl-and ethyl-mercury in sub-nanogram amounts from natural waters on to sulphydryl cotton fibre. Sulphydryl cotton fibre was prepared by soaking cotton for 4 to 5 days in a mixture of thioglycolic acid, acetic anhydride, acetic acid, sulphuric acid, and water at 40 to 45 °C. Adsorbed mercury compounds were eluted from the cotton fibre with hydrochloric acid/sodium chloride and extracted with benzene. Using gas chromatography with electron capture detection, the detection limits were 0.04 ng $litre^{-1}$ using a 20 litre sample volume. Precision was approximately 20 per cent. The method was applied to snow and fresh-water samples of varying humic acid content. Methylmercury concentrations varied between 0.09 and 0.22 ng $litre^{-1}$, and recoveries of spiked methylmercury varied between 42 and 68 per cent and were strongly correlated with the concentration of humic substances. Methylmercury concentration in snow was 0.28 ng $litre^{-1}$ and the recovery was 79 per cent.

References

1. Yoshimura, K., Hiraoka, S., and Tarutani, T. (1982). *Analytica Chimica Acta*, *142*, 101.
2. Muzzarelli, R. A. A. and Rochetti, R. (1973). *Analytica Chimica Acta*, *64*, 371.

3. Muzzarelli, R. A. A. (1971). *Analytica Chimica Acta*, *54*, 133.
4. Vanderborght, B.M. and Van Grieken, R.D. (1977). *Analytical Chemistry*, *49*, 311.
5. Riccardo, A. A., Muzzarelli, R. A. A., and Tubertini, O. (1970). *Journal of Chromatography*, *47*, 414.
6. Muzzarelli, R., Rochetti, R., and Marangio, G. (1972). *Journal of Radioanalytical Chemistry*, *10*, 17.
7. Jha, I. N., Iyengor, L., and Rao, A. V. S. P. (1988). *Journal of Environmental Engineering*, *114*, 962.
8. Muzzarelli, R. A. A. and Rocchetti, R. (1974). *Analytica Chimica Acta*, *69*, 35.
9. Muzzarelli, R. A. A. and Sipos, L. G. (1971). *Talanta*, *18*, 853.
10. Ricardo, A. A., Muzzarelli, R. A. A., and Tubertini, O. (1970). *Journal of Chromatography*, *47*, 414.
11. Muzzarelli, R. A. A. and Sipos, L. (1971). *Talanta*, *18*, 853.
12. Marshall, M. A. and Mottola, A. (1985). *Analytical Chemistry*, *57*, 729.
13. Samara, C. and Kouimtzis, T. A. (1987). *Chemosphere*, *16*, 405.
14. Samara, C. and Kouimtzis, T. A. (1985). *Analytica Chimica Acta*, *174*, 305.
15. Genova, N., Crespi, V., Maggi, L., and Brandove, A. (1983). *International Journal of Applied Radiation and Isotopes*, *34*, 757.
16. Willie, S. N., Sturgeon, R. E., and Berman, S. S. (1983). *Analytical Chemistry*, *55*, 981.
17. Terada, K., Matsumuoto, K., and Inaba, T. (1985). *Analytica Chimica Acta*, *170*, 225.
18. Laydon, D. E., Wegscheider, W., and Bodnar, W. (1979). *International Journal of Environmental Analytical Chemistry*, *7*, 85.
19. Allen, E. A., Boardman, H. C., and Plunckett, B. A. (1987). *Analytica Chimica Acta, 196*, 323.
20. Szabo, G., Guczi, J., and Stur, D. (1987). *Journal of Radioanalytical and Nuclear Chemistry*, *111*, 441.
21. Sandell, E. B. (1959). *In Colorimetric Determination of Traces of Metals*, Interscience, New York.
22. Marchart, H. (1984). *Analytica Chimica Acta*, *30*, 11.
23. Willems, G. J. and de Ranter, C. J. (1974). *Analytica Chimica Acta*, *68*, 111.
24. Myasoedova, G. V. and Savvin, S. G. (1982). *Zhur Analytica Khim*, *37*, 499.
25. Leyden, D. E. and Wegscheider, W. (1981). *Analytical Chemistry*, *63*, 1059A.
26. Willie, S. N., Sturgeon, R. E., and Berman, S. S. (1983). *Analytical Chemistry*, *55*, 981.
27. Sturgeon, R. E., Berman, S. S., Willie, S. N., and Desauliniers, J. A. H. (1981). *Analytical Chemistry*, *53*, 2337.
28. Terada, K., Morimoto, K., and Kiba, T. (1980). *Analytica Chimica Acta*, *116*, 127.
29. Volkan, M., Ataman, O. Y., and Howard, A. G. (1987). *Analyst (London)*, *112*, 1409.

30. Nakashima, S., Sturgeon, R. E., Willie, S. N., and Berman, S. S. (1988). *Fresenius Zeitschrift Fur Analytische Chemie*, *330*, 592.
31. Leydon, D. E. and Underwood, A. L. (1964). *Journal of Physical Chemistry*, *68*, 2093.
32. Florence, J. M. and Batley, G. E. (1976). *Talanta*, *23*, 179.
33. Riley, J. P. and Taylor, D. (1968). *Analytica Chimica Acta*, *41*, 175.
34. Abdullah, M. J. and Royle, L. G. (1972). *Analytica Chimica Acta*, *58*, 283.
35. Figura, P. and McDuffle, B. (1977). *Analytical Chemistry*, *49*, 1950.
36. Loewenschuss, H. and Schmuckler, G. (1964). *Talanta*, *11*, 2093.
37. Dowex A-1 Chelating resin. Dow Chemical Company, Midland, Michigan, USA. (1959).
38. Le Meur, J. F. and Courtot-Coupez, J. (1973). *Bulletin of the Chemical Society of France*, 929.
39. Voutsa, D., Samara, K., Fyticaros, K., and Kouimtzis, T. (1988). *Fresenius Zeitschrift fur Analytische Chemie*, *330*, 596.
40. Pakalns, P., Batley, G. E., and Cameron, A. J. (1978). *Analytica Chimica Acta*, *99*, 233.
41. Pakalns, P., Batley, G. E., and Cameron, A. J. (1978). *Analytica Chimica Acta*, *99*, 233.
42. Subramanian, K. S., Meranger, J. C., Langfond, C. H., and Allen, C. (1983). *International Journal of Environmental Analytical Chemistry*, *16*, 33.
43. Riley, J. P. and Taylor, D. (1968). *Analytical Chemistry*, *41*, 175.
44. Clanet, F., Delangele, R., and Popoff, G. (1981). *Water Research*, *15*, 591.
45. Everaets, F. M., Verbeggen, M., Reijenga, J. C., Aben, G. U. A., Gebauer, P. E., and Bocek, P. (1985). *Journal of Chromatography*, *320*, 263.
46. Pakalns, P. (1980). *Analytica Chimica Acta*, *120*, 289.
47. Gladney, E. S., Peters, R. J., and Perrin, D. R. (1983). *Analytical Chemistry*, *55*, 976.
48. Hinse, A., Kabori, K., and Ishii, D. (1978). *Analytica Chimica Acta*, *97*, 303.
49. Greenberg, R. R. and Kingston, H. M. (1983). *Analytical Chemistry*, *55*, 1160.
50. Van Grieken, R. E., Bresseleers, C. M., and Vanderborght, B. M. (1977). *Analytical Chemistry*, *49*, 1326.
51. Van Espen, P., Nullers, H., and Adams, F. C. (1977). *Zeitschrift fur Analytische Chemie*, *285*, 215.
52. Figura, P. and McDuffie, B. (1980). *Analytical Chemistry*, *52*, 1433.
53. Minagawa, K., Takazawa, Y., and Kufune, I. (1980). *Analytica Chimica Acta*, *115*, 1103.
54. Yamagami, E., Tateishi, S., and Hashimoto, A. (1980). *Analyst (London)*, *105*, 491.
55. Zlatkis, A., Bruening, W., and Bayer, E. (1971). *Analytica Chimica Acta*, *56*, 399.
56. Samchuk, A. I., Kazakevich, Y. E., Romanov, N. N., Danilova, E. Y.,

Khabazova, T. A., and Fedotov, K. V. (1988). *Soviet Journal of Water Chemistry and Technology*, *10*, 63.
57. Imai, S., Muroi, M., Hamaguchi, A., and Kuyama, H. (1983). *Analytical Chemistry*, *55*, 1215.
58. Smits, J. and Van Grieken, R. (1981). *International Journal of Environmental Analytical Chemistry*, *9*, 81.
59. Reggeks, G. and Van Grieken, R. (1984). *Fresenius Zeitschrift fur Analytische Chemie*, *317*, 520.
60. Burba, P., Willmer, P. G., Betz, M., and Fuchs, S. (1983). *International Journal of Environmental Analytical Chemistry*, *13*, 177.
61. Burba, P., Rober, M., and Lieser, K. H. (1978). *Angew. Makromolekulare. Chemistry*, *66*, 131.
62. Runnels, J. H., Merryfield, R., and Fisher, H. B. (1975). *Analytical Chemistry*, *47*, 1258.
63. Burba, P. and Willmer, P. G. (1982). *Analytical Chemistry*, *311*, 222.
64. Brajster, K. and Slonawka, B. (1986). *Analytica Chimica Acta*, *185*, 271.
65. Burba, P., Cebulo, M., and Broekaert, J. A. C. (1984). *Fresenius Zeitschrift fur Analytische Chemie*, *318*, 1.
66. Burba, P. and Willman, P. G. (1983). *Talanta*, *30*, 381.
67. Masri, M. S. and Friedman, M. (1972). *Environmental Science and Technology*, *6*, 745.
68. Goldbach, K. and Lieser, K. H. (1982). *Fresenius Zeitschrift fur Analytische Chemie*, *311*, 183.
69. Burba, P. and Willmer, P. G. (1986). *Fresenius Zeitschrift fur Analytische Chemie*, *324*, 298.
70. Pai, S. C., Whung, P. Y., and Lai, R. L. (1988). *Analytica Chimica Acta*, *211*, 257.
71. Pai, S.C. (1988). *Analytica Chimica Acta*, *211*, 271.
72. Kingston, H. M., Barnes, I. L., Brady, T. J., Rains, T. C., and Champ, M. A. (1978). *Analytical Chemistry*, *50*, 2064.
73. Bruland, K. W., Franks, R. P., Knauer, G. A., and Martin, J. G. (1979). *Analytica Chimica Acta*, *105*, 233.
74. Sturgeon, R. E., Bermann, S. S., Desauliniers, A., and Russell, D. S. (1980). *Talanta*, *27*, 85.
75. Rasmussen, L. (1981). *Analytica Chimica Acta*, *125*, 117.
76. Greenberg, R. R. and Kingston H. M. (1982). *Journal of Radioanalytical Chemistry*, *71*, 147.
77. Paulson, A. J. (1986). *Analytical Chemistry*, *58*, 183.
78. Kingston, H. M., Barnes, I. L., and Brady, T. J. (1978). *Analytical Chemistry*, *50*, 2064.
79. Bender, M. L., Klinkhammer, G. P., and Spencer, D. W. (1977). *Deep Sea Research*, *24*, 799.
80. Klinkhammer, G. P. and Bender, M. L. (1980). *Earth Planet Science Letters*, *46*, 361.
81. Batley, G. E. and Farrah, Y. J. (1978). *Analytica Chimica Acta*, *99*, 283.
82. Gardner, J. and Yates, J. (1979). In *International Conference, Management and Control of Heavy Metals in the Environment*, Water Research

Centre, Stevenage Laboratory, Stevenage, UK; CEP Consultants, Edinburgh, pp. 427–430.
83. Bonforti, R., Ferraroli, R., Frigieri, P., Heltai, D., and Queirezza, G. (1984). *Analytica Chimica Acta*, *162*, 33.
84. Olsen, S., Pessanda, L. C. R., Ruzicka, J., and Hansen, E. H. (1983). *Analyst (London)*, *108*, 905.
85. Jorgensen, S. S. and Petersen, K. (1983). In *Paper presented at 9th Nordic Atomic Spectroscopy and Trace Element Conference*, Reykjavik, Iceland.
86. Fang, S., Xu, L., and Zhang, S. (1984). *Analytica Chimica Acta*, *164*, 41.
87. Malamas, F., Bengtsson, M., and Johansson, G. (1984). *Analytica Chimica Acta*, *169*, 1.
88. Krug, F. J., Reis, B. F., and Jorgensen, S. S. (1983). In, *Proceedings of the Workshop on Locally Produced Laboratory Equipment for Chemical Education*. Copenhagen, Denmark. p. 121.
89. Fang, Z., Ruzicka, J., and Hansen, E. H. (1984). *Analytica Chimica Acta*, *164*, 23.
90. Berman, S. S., McLaren, J. W., and Willie, S. N. (1980). *Analytical Chemistry*, *52*, 488.
91. Sturgeon, R. E., Berman, S. S., and Desauliniers, J. A. U. (1980). *Analytical Chemistry*, *52*, 1585.
92. Kingston, H. M., Barnes, I. L., and Brady, T. J. (1978). *Analytical Chemistry*, *50*, 2064.
93. Greenberg, R. R. and Kingston, H. M. (1982). *Journal of Radioanalytical Chemistry*, *71*, 147.
94. Lee, C., Kim, N. P., Lee, I. C. and Chung, K. S. (1977). *Talanta*, *24*, 241.
95. Piper, D. S. and Goles, G. G. (1969). *Analytica Chimica Acta*, *47*, 560.
96. Ghoda, S. (1972). *Bulletin of the Chemical Society of Japan*, *45*, 1704.
97. Brine, E. (1972). *Report Aktiebolaget Atomenerai*, AE-466.
98. Fujinaga, T., Kusaka, R., Koyama, M. (1973). *Journal of Radioanalytical Chemistry*, *13*, 301.
99. Lieser, K. H., Calmano, W., Heuss, E., and Neitzert, V. (1977). *Journal of Radioanalytical Chemistry*, *37*, 717.
100. Lee, C., Kim, N. B., Lee, I. C., and Chung, K. S. (1977). *Talanta*, *24*, 241.
101. Lo, J. M., Wei, J. C., and Yeh, S. J. (1977). *Analytical Chemistry*, *49*, 1146.
102. Greenberg, R. R. and Kingston, H. M. (1982). *Journal of Radioanalytical Chemistry*, *71*, 147.
103. Greenberg, R. R. and Kingston, H. M. (1983). *Analytical Chemistry*, *55*, 1160.
104. Murthey, R. S. S. and Ryan, D. E. (1983). *Analytical Chemistry*, *55*, 682.
105. Huhn, C. A. and Bacon, M. P. (1985). *Analytical Chemistry*, *57*, 2065.
106. Stiller, M., Mantel, M., and Rappaport, M. S. (1984). *Journal of Radioanalytical and Nuclear Chemistry*, *83*, 345.
107. Matthews, A. D. and Riley, J. P. (1970). *Analytica Chimica Acta*, *51*, 287.
108. Isshiki, K. and Nakayama, E. (1987). *Analytical Chemistry*, *59*, 291.
109. Mykytiuk, A. P., Russell, D. S., and Sturgeon, R. E. (1980). *Analytical Chemistry*, *52*, 1281.

110. Sturgeon, R. E., Berman, S. S., Desauliniers, A., and Russell, D. S. (1980). *Talanta*, *27*, 85.
111. Blake, W. E., Bryant, M. W. R., and Waters, A. (1969). *Analyst (London)*, *94*, 49.
112. Nakahara, T. and Chakrabarti, C. L. (1979). *Analytica Chimica Acta*, *104*, 99.
113. Van den Sloot, H. A., Wals, G. D., and Das, H. A. (1977). *Progress in Water Technology*, *8*, 193.
114. Van den Sloot, H. A. and Das, H. A. (1977). *Progress in Water Technology*, *8*, 193.
115. Muzzarelli, R. A. A. and Rochetti, R. (1973). *Analytica Chimica Acta*, *64*, 371.
116. Murthy, R. S. S. and Ryan, D. E. (1982). *Analytica Chimica Acta*, *140*, 163.
117. Knockel, A. and Prange, A. (1981). *Fresenius Zeitschrift fur Analytische Chemie*, *306*, 252.
118. Omar, M. and Bowen, H. J. M. (1982). *Analyst (London)*, *107*, 654.
119. Maloney, M. P., Moody, G. J., and Thomas, J. D. R. (1980). *Analyst (London)*, *105*, 1087.
120. Braun, T. and Abbas, M. N. (1980). *Analytica Chimica Acta*, *119*, 113.
121. Torak, S., Braun, Pl, Van Dyck, P., and Van Grieken, R. (1986). *X-Ray Spectrometry*, *15*, 7.
122. Gawargious, Y. A., Abbas, M. N., and Hassan, H. N. A. (1988). *Analytical Letters (London)*, *21*, 1477.
123. Testa, C. and Staccioli, L. (1972). *Analyst (London)*, *97*, 527.
124. Delle Site, A., Marchionni, V., and Testa, C. (1980). *Analytica Chimica Acta*, *117*, 217.
125. Testa, C. and Delle Site, A. (1976). *Journal of Radioanalytical Chemistry*, *34*, 121.
126. Horvath, Z. and Barnes, R. M. (1986). *Analytical Chemistry*, *58*, 1352.
127. Dingman, J., Siggia, S., Barton, C., and Hiscock, K. B. (1972). *Analytical Chemistry*, *44*, 1351.
128. Fitzgerald, W. F., Lyons, W. B., and Hunt, C. D. (1974). *Analytical Chemistry*, *46*, 1882.
129. Kramer, H. J. and Neidhart, J. (1977). *Journal of Radioanalytical Chemistry*, *37*, 836.
130. De Mora, S. J. and Harrison, R. M. (1983). *Analytica Chimica Acta*, *153*, 307.
131. *Official Journal of the European Community*, (1980). No. L229/11(80/778/EEC), 23.
132. Corsini, A., Chiang, S., and Di Frucia, R. (1982). *Analytical Chemistry*, *545*, 1433.
133. Wan, C., Chiang, S., and Corsini, A. (1985). *Analytical Chemistry*, *57*, 719.
134. Taguchi, S., Yamazaki, S., Yamamoto, A., Urayaga, Y., Hata, N., Kasahara, I., and Goto, K. (1988). *Analyst (London)*, *113*, 1695.
135. Ohzeki, K., Uno, T., Nukatsuku, I., and Ishida, R. (1988). *Analyst (London)*, *113*, 1545.

136. Mundschenk, H. (1979). *Gewasserkundliche Mitteilungen*, *23*, 64.
137. Yoshimura, K., Kariya, R., and Torntani, T. (1979). *Analytica Chimica Acta*, *109*, 115.
138. Taguchi, S., Ito-oko, E., Matsuyama, K., and Kashara, I. Gotok. (1985). *Talanta*, *32*, 391.
139. Smits, J. and Van Grieken, R. (1981). *Analytica Chimica Acta*, *123*, 9.
140. Smits, J. and Van Grieken, R. (1981). *International Journal of Enviromental Analytical Chemistry*, *9*, 81.
141. Farag, A. G., El-Wakl, A. M., and El-Shahawi, M. S. (1981). *Analyst (London)*, *106*, 809.
142. Khan, A. S. and Chow, A. (1963). *Analytical Letters (London)*, *16*, 265.
143. Palagni, S. (1983). *International Journal of Applied Radiation and Isotopes*, *34*, 755.
144. Tombo, N., Kameri, T., Nishimura, T., and Fukushi, K. (1979). *Japan Water Works Association* No. 532, 37.
145. Hiraide, E. M., Tillkeratue, K., Otsuka, K., and Mizuike, A. (1985). *Analytica Chimica Acta*, *172*, 215.
146. Hine, P. T. and Bursill, D. B. (1984). *Water Research*, *18*, 1461.
147. Hezzlar, J. (1987). *Water Research*, *21*, 1311.
148. Minear, R. A. (1972). *Environmental Science and Technology*, *6*, 431.
149. Minear, R. A. and Walonski, K. I. (1974). In *Investigation of the Chemical Identity of Soluble organophosphorus compounds found in natural waters Research Report*, UILUWRC-74-0086.
150. Ahling, B. and Jensen, S. (1970). *Analytical Chemistry*, *42*, 1483.
151. Kurtz, D. A. (1977). In *Bulletin of Environmental Contamination and Toxicology*, *17*, 391.
152. Musty, P. R. and Nickless, G. (1974). *Journal of Chromatography*, *100*, 83.
153. Ahmed, S. M., Beasley, M. D., Etromson, A. C., and Hites, R. A. (1974). *Analytical Chemistry*, *46*, 1858.
154. Schatzberg, P. and Jackson, D. F. (1972). In *US Coast Guard Report No. 734209.9*, Washington DC. November.
155. Carsin, J. L. (1977). *Revue Internationale d'Oceanographic Medicale*, *48*, 77.
156. Gesser, H. D., Chow, A., Davis, F. C., Uthe, J. F., and Reinke, J. (1971). *Analytical Letters (London)*, *4*, 883.
157. Bedford, J. W., (1974). *Bulletin of Environmental Contamination and Toxicology*, *12*, 662.
158. Gesser, H. D., Chow, A., Davis, F. C., Uthe, J. F., and Reinke, J. (1971). *Analytical Letters (London)*, *4*, 883.
159. Uthe, J. F., Reinke, J., and Gesser, H. (1972). *Environmental Letters*, *3*, 117.
160. Saxena, J., Basu, D. K., and Kozuchowski, J. (1977). In *Report No. PB276635*, US National Technical Information Service, Springfield, Va, p. 94.
161. Saxena, J., Kozuchowski, J., and Basu, D. K. (1977). *Journal of Environmental Science and Technology*, *11*, 682.

162. Basu, D. K. and Saxena, J. (1978). *Journal of Environmental Science and Technology*, *12*, 791.
163. Fox, A. and Staley, A. M. (1976). *Analytical Chemistry*, *48*, 992.
164. Navra'til, J. D., Sievers, R. E., and Walton, H. F. (1977). *Analytical Chemistry*, *49*, 2260.
165. Afghan, B. K., Wilkinson, R. J., Chow, A., Findley, T. W., Gesser, H. D., and Srikameswaren, K. S. (1984). *Water Research*, *18*, 9.
166. Weil, L. (1972). *Gas-u Wassfach*, *113*, 64.
167. Weil, L. (1973). *Analytical Abstracts*, *24*, 1259.
168. Beyermann, K. and Eckrich, W. (1974). *Zcit Analytical Chemistry*, *265*, 1.
169. Rice, M. R. and Gold, H. S. (1984). *Analytical Chemistry*, *56*, 1436.
170. Josefson, C. M., Johnston, J. B., and Trubey, R. (1984). *Analytical Chemistry*, *56*, 764.
171. Aue, W. A., Kapila, S., and Hastings, C. R. (1972). *Journal of Chromatography*, *73*, 99.
172. Kawabata, N. and Chira, K. (1979). *Environmental Science and Technology*, *13*, 1396.
173. Carpenter, A., Siggia, S., and Carter, S. (1976). *Analytical Chemistry*, *48*, 225.
174. O'Keefe, P., Meyer, C., Smith, R., Hilker, D., Aldous, K., and Wilson, L. (1986). *Chemosphere*, *15*, 1127.
175. Stuber, H. and Leenheer, J. A. (1983). *Analytical Chemistry*, *55*, 111.
176. Grob, R. L. and Kaiser, O. (1976). *Journal of Environmental Science and Health*, *A11*, 623.
177. Robinson, J. L., John, J., Safa, A. I., Kirkes, K. A., and Griffith, P. E. (1987). *Journal of Chromatography*, *402*, 201.
178. Gawdzik, J. and Gawdzik, B. (1988). *Czerwinska—Bil U. Chromatographia*, *25*, 504.
179. Noren, K. and Sjovall, J. (1987). *Journal of Chromatography*, *414*, 55.
180. Dedek, W., Wenzel, K. D., Luft, F., Oberlander, H., and Mothes, B. (1987). *Fresenius Zeitschrift fur Analytische Chemie*, *328*, 484.
181. Musty, P. R. and Nickless, G. (1976). *Journal of Chromatography*, *120*, 369.
182. Braun, T., Abbas, M. N., Torak, S., and Zvakefalvi-Nagy, Z. (1984). *Analytica Chimica Acta*, *160*, 277.
183. Braun, T., Abbas, M. M., Bakos, L., and Elek, A. (1981). *Analytica Chimica Acta*, *131*, 311.
184. Lee, Y. H. (1987). *International Journal of Environmental Analytical Chemistry*, *29*, 263.

2

METAL OXIDES AND METAL ADSORBENTS

2.1 Metal cations

2.1.1 Manganese dioxide

Non-saline waters Manganese dioxide supported on glass fibres has been used for the preconcentration of lead from natural waters.[1] Up to 75 mg of lead are adsorbed per g of manganese dioxide.

Reagents/materials:

Manganese dioxide: BDH precipitated laboratory reagent. This is in the form of a very fine powder and it was used directly, with no pretreatment.

Filters:

Whatman GF/A 11 cm diameter.

Procedure:

Equipment was designed to collect lead from both aqueous and suspended phases in a single operation. Prepare filters and use them in the filter holder by the following procedure. Place a GF/A filter on the bottom plate and place the collar on the filter. Pour the manganese dioxide as a suspension of 1 g powder in 50–100 ml water on to the filter and apply suction, thus leaving a uniform deposit (approximately 90 mm diameter) of manganese dioxide on the filter. Remove the collar and place two GF/A filters on top of the manganese dioxide making three filter layers with manganese dioxide sandwiched between the bottom two. The top filter retained particulate matter from the water and separated it from the manganese dioxide which adsorbed lead from solution.

Replace the collar and fill with water (approximately 300 ml) then bolt the top plate into position and connect to the liquid reservoir (the liquid was drawn from either 10-litre beakers or 25-litre plastic containers). Then turn on the pump, regulate the flow rate by adjusting pump vacuum (normally set at 1 litre min^{-1}). The volume of water maintained in the collar above the filters ensured uniform flow through the filters during sampling. After sampling, separate three filters and analyse for lead as follows.

Recover lead by heating both the two lower filters in 25–50 ml 6 M hydrochloric acid to dissolve the manganese dioxide. Either directly analyse this solution for lead by atomic absorption spectrophotometry (using the 217 nm line in an air/acetylene flame) or recover it gravimetrically as follows. Filter the manganese dioxide solution and boil (just) to dryness. Add 20 ml water and boil to dryness (to remove remaining hydrochloric acid). Add 50 ml water and heat to dissolve the lead chloride. Then add sulphuric acid and filter the resulting lead sulphate. Dissolve lead sulphate in concentrated ammonium acetate solution and filter. Then add sodium chromate and determine the lead gravimetrically as lead chromate. Boil the uppermost filter of the three, containing filtered material in 2 M nitric acid for 30 min to extract lead and analyse the resulting solution directly by atomic absorption spectrometry using a sample flow rate of 6.5 ml min^{-1}.

In actual river water analysis there was no pH adjustment or other pretreatment of the sampie. The presence of relatively high concentrations of manganese during lead analyses caused no interferences or aspirating problems.

The adsorption capacity of the manganese dioxide was investigated by determining the amount of lead adsorbed from varying amounts in 2-litre potable water solutions. The results are shown in Fig. 6 of Volume 1, curve A, where it can be seen that adsorption is quantitative up to 75 mg lead, with a maximum adsorption of approximately 190 mg Pb g^{-1} lead from 500 mg in solution. Identical behaviour was found when sea-water was substituted for potable water. The importance of the thickness of the manganese dioxide bed was investigated by supporting manganese dioxide on a filter held in a 47 mm diameter Millipore filter holder so that the manganese dioxide disc had a smaller diameter (34 mm) and a thickness of approximately 2 cm or seven times the thickness of the deposit on 11 cm filters. Under these conditions quantitative adsorption of up to approximately 160 mg lead occurred (Fig. 6 of Volume 1, curve B) with a maximum of 300 mg g^{-1} manganese dioxide at a flow rate of 500 ml min^{-1}.

There was no significant difference in lead recovery (minimum 97 per cent) between results obtained in the range pH 3–5.5 with a slight decrease at pH 2. Solutions of higher pH caused difficulties because of the tendency of lead compounds to precipitate. With river water at its natural pH of 7.4, spiked 2 mg lead in a 10 litre sample, 0.6 mg lead was retained on the top particulate filter and 1.4 mg was retained by the manganese dioxide. When a similar 10 litre sample was acidified to pH 3 before extraction of the added lead, only 0.04 mg lead was retained as particulate matter (i.e. 4 μg $litre^{-1}$) similar to the natural level described below, with approximately 2 mg being retained by the manganese dioxide. It seems therefore that changes in pH of the aqueous medium (at least in the range pH 2–7.4) have little effect on the ability of manganese dioxide to adsorb lead, or on the total amount of lead recovered as particulate matter and from solution, but the distribution of lead between suspended and aqueous phases is pH dependent.

Samples of river water collected before and after a heavy rainfall give the following results: before rain—$Pb_{(aq)}$ 1.6 μg litre^{-1}, Pb solid, 5.6 μg litre^{-1}; after rain—$Pb_{(aq)}$ 6.3 μg litre^{-1}, Pb solid 19.8 μg litre^{-1}. The increase after rain was presumably due to lead-bearing run-off from city streets which drain into the river.

Hashimoto *et al.*[2] preconcentrated lead-214 and bismith-214 on Millipore filters, preimpregnated with manganese dioxide, lead-214 and bismith-214 were then measured in the filter material by direct alpha spectrometric analysis. Michel *et al.*[3] preconcentrated radium-226 and radium-228 from 1-litre samples of water on to manganese dioxide impregnated on to acrylic fibres. Radium was then leached from the fibres and coprecipitated with barium sulphate prior to γ counting. Elsinger *et al.*[4] used a similar technique employing 100-litre samples. More recently, Butts *et al.*[5] and Rama *et al.*[6] respectively, have preconcentrated and detemined radium-226 and radium-224 in natural waters by a similar technique.

Sea-water Harvey and Dutton[7] determined nanogram amounts of cobalt in sea-water after preconcentration on manganese dioxide formed by photochemical oxidation of divalent manganese in a photochemical reactor. The sample (1 litre) containing 100 μg manganese was irradiated in a Hanovia reactor fitted with a 2W low-pressure Hg discharge lamp radiating mainly at 254 and 185 nm. The manganese dioxide deposit that adhered to the silica jacket of the reactor was dissolved in 0.15 M hydrochloric acid containing a trace of sulphur dioxide, the solution was evaporated to dryness and the residue was dissolved in 4 ml or 0.625 M hydrochloric acid; 1 ml 5 M aqueous ammonia and 0.1 ml of 0.1 per cent dimethylgloxime in ethanol were added, and cobalt was determined by pulse polarography. The polarograph was operated in the derivative mode, starting at -1.0 V; a 50 mV pulse height and 1 s mercury drop life were used.

Bacon and Anderson[8] preconcentrated thorium-234, thorium-230, and thorium-228 in sea-water on manganese dioxide coated on plastic netting prior to standard radiochemical counting techniques.

Oceanographers have developed methods to measure the radium-226 content of sea-water as it is a useful tracer of mixing in the ocean. These procedures are based on concentrating radium from a large volume of sea-water, removing all thorium-228 from the sample and ageing the sample while a new generation of thorium-228 partially equilibrates with radium-228. After storage periods of 6–12 months, the sample is spiked with thorium-230 and after ion-exchange and solvent exchange separations, the thorium isotopes are measured in a γ-ray spectrometer system utilizing a silicon surface barrier detector. Early work was based on concentrating the radium from the sea-water sample by adding barium and coprecipitating with barium sulphate. This concentration procedure has been replaced by one involving the extraction of radium from sea-water on acrylic fibre coated with manganese dioxide[9,10] (Mn fibres). By use of this technique, volumes of 200–2000 litres may be sampled routinely.

Radium is extracted from ground waters with even better efficiency on the manganese fibre.[11] Radium is removed from the manganese fibre by reducing the manganese dioxide with hot hydrochloric acid or desorbing the radium into cold dilute nitric acid. The hydrochloric acid treatment removes the radium quantitatively from the fibres but generates a considerable quantity of chlorine. The nitric acid treatment is much easier and safer but only removes about 70 per cent of the radium.[10] Measurements of radium-226 are simpler than those for radium-228 and are more precise. These measurements are generally made by concentrating the radium from up to a few litres via barium sulphate precipitation followed by thick source counting or by radon-222 extraction following dissolution of barium sulphate.[12]

Oceanographers use different techniques for measuring radium-226 in sea-water. Some workers store the sample in a 20 litre glass bottle and extract successive generations of radon-222.[13,14] Others quantitatively extract the radium on to manganese fibre and measure radon-222 directly emanating from the manganese fibre[15] or in a hydrochloric acid extract from the fibres.[16] The radon-222 activity is then determined by a α-scintillation counting. All of these techniques give high levels of reproducibility and accuracy as determined by the oceanographic consistency of the results.[13,14]

The introduction of high-resolution, high-efficiency γ-ray detectors composed of lithium-drifted germanium crystals has revolutionized γ-measurement techniques. Thus, γ-spectrometry allows the rapid measurement of relatively low-activity samples without complex analytical preparations. A technique described by Michel *et al.*[17] uses Ge(Li) γ-ray detectors for the simultaneous measurements of radium-228 and radium-226 in natural waters. This method simplifies the analytical procedures and reduces the labour while improving the precision, accuracy, and detection limits. In this method the radium isotopes are preconcentrated in the field from 100 to 1000 litre water sample on to manganese impregnated acrylic fibre cartridges, leached from the fibre and coprecipitated with barium sulphate. Lower limits of detection are controlled by the volume of water processed through the manganese fibres. In a 1 day count, samples as low as 10 dpm are measured to ± 10

Table 26 Comparison of the known values of the EPA radium in water cross-check samples with both the ^{228}Th ingrowth/^{222}Rn emanation and Ge(Li) γ-ray methods.[a]

	Ra/^{228}Ra activity ratio		^{228}Ra (decays min^{-1})	
EPA no.	Th ingrowth	Known	Th ingrowth	Known
Dec 78	1.04 ± 0.06	0.96 ± 0.20	21.1 ± 0.8	19.8 ± 2.9
Mar 79	1.59 ± 0.31	1.66 ± 0.34	28.6 ± 5.5	30.2 ± 4.4
Apr 79	1.05 ± 0.05	1.05 ± 0.22	13.6 ± 0.5	13.8 ± 2.0
	Ge(Li)	Known	Ge(Li)	Known
Sept 80	1.07 ± 0.05	1.10 ± 0.24	29.1 ± 1.5	31.9 ± 1.8
Oct 80	0.66 ± 0.03	0.72 ± 0.09	19.8 ± 1.0	20.4 ± 1.8
Dec 80	0.76 ± 0.06	0.79 ± 0.17	22.0 ± 1.1	23.3 ± 3.6

[a]Measurements are reported as weighted means with weighted standard deviations. The known values are reported by EPA with expected laboratory standard deviations.
From Michel *et al.*[17] with permission.

per cent uncertainty. This manganese fibre–γ-ray technique is shown to be more accurate than the actinium-228 methods recommended by the US Environmental Protection Agency[18] and as accurate but more rapid than the thorium-228 ingrowth procedure.

Table 26 shows the results for samples from the Environmental Protection Agency Radium in Water Crosscheck Program using both the thorium-228 ingrowth and the Ge(Li) γ-ray techniques. Both types of analyses were run on 3.8-litre samples. The thorium-228 ingrowth

Table 27 Comparison of ^{226}Ra, ^{228}Ra, and ^{228}Ra/^{226}Ra activity ratio in sea-water of the same sampling using two different methods.

Sample	^{226}Ra (decays min^{-1})[a]	^{228}Ra (decays min^{-1})[a]	^{228}Ra/^{226}Ra	Method[a]
451–99	260 ± 7	210 ± 10	0.81 ± 0.04	1
	305 ± 2	223 ± 5	0.73 ± 0.02	2
452–99	235 ± 10	125 ± 13	0.53 ± 0.06	1
	229 ± 12	153 ± 3	0.67 ± 0.04	2
453–99	154 ± 1	56 ± 7	0.36 ± 0.04	1
	161 ± 9	53 ± 2	0.33 ± 0.02	2
454–99	246 ± 13	94 ± 11	0.38 ± 0.05	1
	291 ± 8	94 ± 3	0.32 ± 0.01	2

[a]Method 1, Ge(Li) γ-ray spectrometry. Results reported as weighted means and weighted standard deviations of both ISD counting and efficiency uncertainties. Method 2, ^{226}Ra by Rn emanation; ^{228}Ra by ^{228}Th ingrowth. Results reported as ISD counting uncertainties.
From Michel *et al.*[17] with permission.

samples were stored for 3–11 months to accumulate sufficient activity and then counted by α-spectrometry for 2 days. The Ge(Li) samples were stored for 3 weeks and counted for 400–1000 min. Both techniques required a separate analysis for the absolute radium-226 activity whereas the thorium-228 ingrowth method requires an additional analysis for the total radium-226 in the large volume sample to obtain the ^{228}Ra/^{226}Ra activity ratio.

Table 27 shows the results for sea-water samples collected by the manganese dioxide fibre technique and analysed by both the Ge(Li) and thorium-228 ingrowth techniques. The ground-water samples were analysed first by-γ-ray counting followed by fusion of the barium sulphate and the thorium-228 ingrowth and radon emanation techniques. The sea-water samples were analysed in the reverse order. Although the bulk activities of each isotope vary for the different techniques, the radium-228/radium-226 activity ratios are in close agreement.

2.1.2 Silica

Non-saline waters Najiri *et al.*[19] give details of a silica gel preconcentration procedure for the simultaneous determination of trace metals in natural water. It involves preliminary complexation of the metals with 8-hydroxyquinoline, adsorption of the complexes on a chemically bonded silica gel disposable plastic column, elution with methanol, digestion with nitric acid, and determination by inductively coupled plasma emission spectrometry.

Sea-water Prange *et al.*[20] carried out multi-element determinations of dissolved heavy metals in Baltic sea-water by total reflection X-ray fluorescence spectrometry. The metals were separated by chelation adsorption of the metal complexes on lipophilized silica-gel carrier and subsequent elution of the chelates by a chloroform/methanol mixture. Trace element loss or contamination could be controlled because of the relatively simple sample preparation. Aliquots of the eluate were then dispensed in highly polished quartz sample carriers and evaporated to thin films for spectrometric measurement. Recoveries, detection limits, and reproducibilities of the method for several metals were satisfactory.

2.1.3 C18-bonded silica (Carbon-18 Sep-pak)

This preconcentration resin can be used in one of two ways. The method most commonly used is to convert the cations to a chelating complex such as ammonium pyrrolidinedithiocarbamates and then pass the solution down a C18-bonded silica column which retains the complex. The concentrated cations complexes are then desorbed with a small volume of aqueous or aqueous organic solvent. In the second method the C18-bonded silica is loaded with a chelating agent and the solution of cations passed through. Metal chelates are formed and retained on the column.

Saline waters Sturgeon *et al.*[21] preconcentrated antimony(III) and antimony(V) from coastal and sea-waters by adsorption of their ammonium pyrrolidinedithiocarbamate chelates on to C18-bonded silica prior to determination by graphite furnace atomic absorption spectrometry. A detection limit of 0.05 μg litre^{-1} was achieved. Sturgeon *et al.*[21] preconcentrated selenium(IV) by adsorption of their ammonium pyrolidinediethyldithiocarbamate chelates on to C18-bonded silica prior to desorption and determination by graphite furnace atomic adsorption spectrometry. The detection limit was 7 ng litre^{-1} selenium(IV), based on a 300 ml water sample.

Heavy metals determinations in natural waters have been carried out by a procedure involving the simultaneous formation of metal dithiocarbamates and on-line preconcentration using a C18-bonded silica precolumn followed by reversed phase high performance liquid chromatographic separation.[22] The C18 precolumn was previously loaded with cetyltrimethylammonium bromide (cetrimide)—dithiocarbamate ion pair. Metal dithiocarbamates formed and retained on the precolumn were eluted directly to the analytical column with a gradient of acetonitrile and water containing cetrimide (10 mM and 2 mM respectively) buffered to pH 6.8 (with phosphate buffer, 10 mM), separated, and detected by a UV photodiode array detector. This procedure allowed determination of subnanogram quantities of metal ions, including copper(II), and mercury(II) ions in potable water and of cadmium(II), lead(II), cobalt(II), nickel(II), and bismuth(II).

Sea-water Sturgeon *et al.*[21,23] proconcentrated selenium and antimony from sea-water on C18-bonded silica gel prior to determination by graphite furnace atomic absorption spectrometry. The method was based on the complexation of selenium(IV), antimony(III), and antimony(V) with ammonium pyrrolidinedithiocarbamate. These complexes were adsorbed on to a column of C18-bonded silica gel, then eluted with methanol, followed by evaporation to near dryness. The residue was taken up in 1 per cent nitric acid. Concentration factors of 200 could be obtained. Detection limits for selenium(IV), antimony(III), and antimony(V) were 7, 50, and 50 ng litre^{-1} respectively, based on a 300 ml sample volume.

Sunda and Hanson[24] preconcentrated free cupric ions in sea-water using a ligand competition technique involving copper sorption on to carbon-18 Sep-Pak followed by a competitive titration of naturally complexed copper with standard EDTA solution.

2.1.4 Titanium dioxide, zirconium dioxide, alumina, and hydrated ferric oxide

Non-saline waters Lieser *et al.*[25] showed that heavy metals, iron, cobalt, nickel, copper, zinc, lead, and uranium, are bound to titanium dioxide, zirconium dioxide, and alumina in varying degrees. Remarkably high fractions of zinc, lead, and uranium are taken up by titanium dioxide and by aluminium oxide. The comparison between the sorbent fraction and the fraction present in suspended matter shows that these fractions are comparable for iron, cobalt, nickel, and copper, but not for zinc, lead, and uranium.

Zhang *et al.*[26] preconcentrated lead by depositing it on a microcolumn of activated alumina and eluting with nitric acid before determination by flame atomic absorption spectrometry. The limit of detection for a 25 ml sample was 0.36 μg litre^{-1}. The relative standard deviation for concentrations of 40 and 4 μg litre^{-1} were 1.4 and 12 per cent, respectively. Schell *et al.*[27] preconcentrated americium and plutonium from 4000 litres water by passage down a column of 0.3 μm Millipore filters and sorption beds of alumina.

Indium has been preconcentrated from rain-water by adsorption on hydrated ferric oxide which is then irradiated with neutrons.[28] Beta particles from the indium-116 were counted with a gas flow proportional counter supplemented with X-ray counting with a Ge(Li) detector.

2.1.5 Gold, silver, and copper

Non-saline waters Gold has been used for the preconcentration of mercury.[29-39] Neske *et al.*[33] determined mercury at the μg litre^{-1} level ng litre^{-1} level. The method is based on the preconcentration using sample sizes of up to 1 litre, using direct contact with a gold collector followed by high sensitivity atomic absorption spectrometry. Determinations of organomercury compounds were also possible. Wittman[37] heated water samples in an electric furnace and collected the mercury released on a gold wire. The preconcentrated mercury is then vaporized and carried through an absorption cell, so that the mercury atomic absorption sequel at 253.7 nm increases sharply to a maximum. Mercury in water has been preconcentrated from lake-water by reduction to elemental mercury, purging out mercury with an inert gas, and collecting the mercury on a gold trap.[36] Mercury was then released from the trap as a sharp pulse by heating and determined by atmospheric pressure helium microwave induced plasma emission spectrometry.

Kunert *et al.*[39] preconcentrated mercury in potable water on gold or silver foil prior to its determination by cold vapour graphite furnace atomic absorption spectrometry and cold vapour atomic absorption spectrometry. Using adsorption on gold or silver wire Temmerman *et al.*[38] were able to determine down to 6.6 μg litre^{-1} mercury in potable water.

The preconcentration of mercury on copper or silver foil has been studied.[40–44] In this procedure the mercury present in a large volume sample is swept on to the silver or copper and subsequently released by heating to 800 °C. The concentrated pulse of mercury swept with argon into cold vapour atomic absorption spectrometer. Down to 0.1 μg litre^{-1} mercury has been determined this way.

Sea-water Olafsson[45–47] described a procedure for the preconcentration and determination of mercury in sea-water in which the sample (450 ml) is acidified with nitric acid, aqueous stannous chloride is added, and the mercury is entrained in a stream of argon into a silica tube wound externally with resistance wire and containing pieces of gold foil, on which the mercury is retained. The tube and its contents are then heated electrically to about 320 °C and the vaporized mercury is swept by argon into a 10 cm silica absorption-cell in an atomic-absorption spectrophotometer equipped with a recorder. The absorption (measured at 253.7 nm) is directly proportional to the amount of mercury in the range 0–24 ng per sample.

Gill and Fitzgerald[48] isolated mercury from sea-water using tin(II) chloride (0.5 ml of 50 per cent solution) reduction and gas phase stripping with collection and concentration on gold columns. The gas flow system used two gold-coated bead columns (the collection and the analytical columns) to transfer mercury into the gas cell of an atomic absorption spectrometer. By careful control and estimation of the blank, a detection limit of 0.21 pM was achieved using 2 litres of sea-water. The accuracy and precision of this method were checked by comparison with aqueous laboratory and National Bureau of Standards (NBS) reference materials spiked into acidified natural water samples at picomolar levels.

2.2 Anions

2.2.1 C18-bonded silica (C18 Sep-Pak)

Non-saline water Hexavalent chromium (chromate) has been preconcentrated from lake water on to C18-bonded silica prior to analysis by high performance liquid chromatography utilizing an atomic absorption spectrophotometric detector.[49] Natural pond water samples were adjusted to pH 3.2 with acetic acid. The samples were analysed and 100 per cent recovery of chromate was obtained even in the presence of chromium(III) ions. Chromate was eluted from the column (methanol/water 50/50) directly to the AAS detector. Alumina has been used to preconcentrate chromate before determination by inductively coupled plasma atomic emission spectrometry.[50] Trivalent chromium is not adsorbed under these conditions. A detection limit of 0.2 μg $litre^{-1}$ was achieved by this procedure.

A particularly recent innovation that has been used for metals but also for chromate preconcentration consists of a modification of the flow injection analysis technique whereby the samples pass through a microcolumn containing an adsorbant for the ion of interest, thereby achieving a concentration factor. Following an automatic switch to an acidic reagent, the adsorbed anion is desorbed in a sharp pulse of the flowing reagent and then passes on to the detection system. Systems such as this, therefore, combine preconcentration and automation. Syty *et al.*[75] separated chromium (III) from chromium(VI) on a microcolumn of alumina which preconcentrated chromium(VI). Chromium(VI) was then flushed from the column with a small volume of acid before determination by inductively coupled plasma atomic emission spectrometry in amounts down to 0.2 μg $litre^{-1}$.

2.2.2 Cadmium exchanged zeolites

Non-saline waters Desalvo and Street[51] preconcentrated sulphide from well water on a column of cadmium exchanged zeolite sorbent prior to determination by the methylene blue visible spectrophotometric method. This method had an appreciably lower detection limit than conventional sulphide methods.

2.3 Organics

2.3.1 C18-bonded silica (C18 Sep-Pak)

Organochlorine insecticides Bagnati *et al.*[52] give details of equipment and procedure for simultaneous semi-quantitative determination of sub-μg $litre^{-1}$ traces of 21 pesticides in ground-water, involving rapid extraction using C18 silica bonded phase columns and capillary gas chromatography with selected ion-recording mass spectrometry. Sherma[53] preconcentrated organochlorine insecticide residues from natural water using solid phase extraction cartridges. Concentrated extracts were spotted directly on to silica gel TLC plates for separation. Detection with ammoniacal silver nitrate reagent was followed by quantification using densitometric scanning. DDT and methoxychlor recoveries were greater than 90 per cent for samples spiked at 2.5 and 0.25 μg litre^{-1}.

Polychlorinated biphenyls Thome and Vandaele[54] preconcentrated polychlorinated biphenyls using microcolumns by quantitative adsorption on to covalently bonded silica gel with octadecyl groups. Maximal efficiency of adsorption was achieved by successive treatment with 1 ml hexane, acetone, methanol, and methanol in water (1:1 v/v). A sample flow rate of 10 ml min^{-1} was achieved by vacuum which was maintained to dry the column. Intefering polar compounds were eluted with 0.2 ml methanol/water. Polychlorinated biphenyls were eluted with 2 ml hexane, concentrated by evaporation, and analysed by gas chromatography with electron capture detection. Mean recovery was 95.2 per cent with polychlorinated biphenyl concentrations of 0.01–10 ppb.

Herbicides Sep-Pak C-18 has been used for the preconcentration of various organic substances in water. Pyrazon herbicide in natural water was adsorbed from a large volume of water then desorbed into a small volume of methylene dichloride prior to analysis by high performance liquid chromatography.[55]

Sherma[57] described a procedure for preconcentrating and determining triazene and chlorophenoxy acid herbicides from natural waters at 10 μg litre^{-1} concentrations involving extraction and concentration on disposable columns containing C18-bonded silica gel, separation by thin layer chromatography on preadsorbent silica gel layers impregnated with silver nitrate, and detection by exposure to UV light with densitometric scanning. Herbicide concentrations were interpolated from calibration lines relating peak areas to weight of standards spotted on the TLC plate. Recoveries from triazines and chlorophenoxy herbicides were 70–88 per cent and 93–100 per cent, respectively.

West and Turner[58] used Sep-Pak C-18 cartridges to extract fluridone from water samples while *N*-methylformamide passed through unretained. Water containing *N*-dimethylformamide was treated with methanol and glycerol before evaporation *in vacuo*. Residual *N*-dimethylformamide was dissolved in methanol and analysed by gas chromatography with a Hall electrolytic conductivity detector operated in the nitrogen mode. Fluridone was eluted from the cartridge with methanol and analysed by liquid chromatography with UV detection at 313 nm. Limits of detection for fluridone and *N*-dimethylformamide were, respectively, 1 and 2 μg litre^{-1}. Mean fluridone recoveries were 98 per cent at 1–400 μg litre^{-1}, and mean *N*-dimethylformamide recoveries were 87 per cent at 2–50 μg litre^{-1}.

Polyaromatic hydrocarbons Ogan *et al.*[59] preconcentrated polyaromatic hydrocarbons on a column of Sep-Pak prior to desorption and determination by high performance liquid chromatography using a spectrofluorimetric detector. Polyaromatic hydrocarbons in potable water have been preconcentrated on Sep-Pak C-18.[56]

Aliphatic diamines Nishikawa[60] preconcentrated aliphatic diamines (after derivitivization with acetylacetone) on Sep-Pak C-18 cartridges. The diamines were determined by high performance liquid chromatography in the range 1.23–14.32 μg litre^{-1} with relative standard deviations of 0.4–4.9 per cent, and detection limits of 0.14–1.78 μg litre^{-1} were estimated for 100 ml water samples. Analyte recovery was 88–96 per cent in river water (relative standard deviations 1.3–4.0 per cent), and 34–93 per cent in sea-water (relative standard deviation 1.3–9.8 per cent).

Miscellaneous organic compounds Sauer *et al.*[61] studied the use of Sep-Pak C-18 bonded silica cartridges for the preconcentration of organic compounds from estuary and sea-water. Nielen *et al.*[62] used precolumns packed with C18-bonded silica and sulphonic acid cation exchange materials for the on-line separation and trace enrichment of trace organics in industrial waste water samples. The precolumn fractions were analysed by automated liquid chromatography and diode array ultraviolet visible detection. Multisignal plots and three-dimensional spectrochromatograms were used in the identification of the pollutants. The technique was optimized for the analysis of 29 organic compounds in industrial waste water. 2,5-Dimethylphenol, 2,6-dichlorophenol, and *o*-phenylphenol recoveries were all in excess of 90 per cent, when the water sample was first acidified to pH 2 and 30 g sodium chloride per 100 ml added before passage through the column. For phenol, however, a recovery of only 42 per cent was achieved under these conditions.

2.3.2 Poropak

Non-saline waters

Nitro-compounds A method has been described[63] for the preconcentration of nitro-compounds in munitions works effluents by adsorption on Poropak resins. Following desorption with acetone the nitro-compounds (nitramines, nitrotoluenes, and nitroaliphatics) were measured by high performance liquid chromatography at a gold/mercury electrode. The results of analysis of munitions in test samples shows that Poropak resins were superior in performance to RDX resins. Analysis time was approximately 2 h per sample for isolation and quantification. Detection limits of 1 μg litre^{-1} were approached.

2.3.3 *Thermotrap TA*

Non-saline waters Ghaoui[64] compared Thermotrap TA an organic adsorbent, with a variety (C8–C18) of reversed phase adsorbents normally used to preconcentrate polynuclear aromatic hydrocarbons from aqueous samples. An aliquot of aqueous solution containing 0.4 mg litre^{-1} each of fluorene, anthracene, phenanthrene, flouranthene, pyrene, chrysene, and benzo(a)pyrene in 2 per cent aqueous tetrahydrofuran was passed, with suction, through each of the adsorbent cartridges. Thermotrap TA gave the highest recoveries of all phases tested for fluorene (98 per cent), anthracene (92 per cent), phenanthrene (less than 40 per cent), fluoranthene (73 per cent), and pyrene (70 per cent) and better recoveries than C8–C18 for chrysene. However, lower recoveries than the phenyl sorbent were obtained for chrysene and benzo(a)pyrene.

2.3.4 *C8 reversed phase adsorbent*

Non-saline waters Noroozian *et al.*[65] used a simple on-line technique involving sorption on a LC microprecolumn packed with a reversed-phase sorbent (C8) followed by direct elution into a gas chromatograph with hexane, to preconcentrate and determine a series of chlorinated pesticides and polychlorinated biphenyls in aqueous samples. Determinations were made at the ng litre^{-1} level for sample volumes of 1.0 ml. In experiments using chemical standards, recoveries 95 per cent and more were observed for the majority of these compounds.

2.3.5 *Silica gel modified cellulose*

Non-saline waters

Phenols Columns containing silica gel chemically linked to different groups have been compared for their suitability for preconcentration of selected phenols from dilute aqueous solution. The chemically bonded phases comprised octyl, octadecyl, phenyl, diol, and cyanide radicals of which the phenyl bonded silica gel gave the best recoveries following reversed phase high performance liquid chromatography analysis.

2.3.6 *Other inorganic adsorbents*

Non-saline waters Sackmauer *et al.*[66] used silica acid-celite columns to preconcentrate organochlorine insecticides. Aue *et al.*[67] used silicones which were chemically bonded to diatomaceous earth particles to extract non-polar organics from water. Natural levels of nucleic acids in lake waters have been preconcentrated on hydroxygraphite.[68] Molecular sieves have been used for the preconcentration of mineral, animal, and vegetable oils in water,[69] alumino silicates (Partisil ODS) for the preconcentration of substituted phenyl urea herbicides in natural and waste waters,[70] and Zeolite 25 H-5 for the preconcentration of low molecular weight aldehydes as their 2,4-dinitrophenylhydrazones in potable water.[71] In the latter method the 2,4-dinitrophenylhydrazones were then separated by liquid chromatography and determined by gas chromatography using flame ionization detection. Results are presented for the separation and detection of a range of aldehydes and ketones in aqueous solution at concentrations of 1000 mg litre^{-1}. Recoveries approximated to 100 per cent for model compounds tested, with the exception of formaldehyde.

2.4 Organometallic compounds

2.4.1 Nucleosil 10–C18 and Tropolin-Cl8 silica cartridges

Non-saline water

Lead. To preconcentrate trialkyl-lead species Blaszkewicz *et al.*[72] complexed interfering metal ions in rain-water with EDTA before adjustment of the pH to 10. Samples were pumped through an extraction column of silica gel to adborb lead compounds which were then desorbed with acetate buffer containing methanol at pH 3.7. The eluate was diluted and adjusted to pH 8 with borate buffer before further concentration on 10–Cl8 precolumn. Adsorbed trialkyl-lead compounds were eluted by back-flushing on to a RP-Cl8 column and separated with methanolic acetate buffer. On-line detection used a post-column chemical reaction detector. Detection limits for sample volumes of 500 ml were 15 μg litre^{-1} and 20 μg litre^{-1} for trimethyl- and triethyl-lead, respectively. Standard deviation was less than 4 per cent for a sample containing 90 pg triethyl-lead per ml.

Tin. Muller[73] has described a procedure for the preconcentration and determination of mono-, di-, tri-, and tetra-substituted organotin compounds in sewage, sludge, and lake water samples. The ionic compounds were extracted from diluted aqueous solutions as chlorides by using a Tropolin-Cl8 silica cartridge and from sediments and sludges by using an ethereal tropolone solution. The extracted compounds were then ethylated by a Grignard reagent, and analysed by high-resolution gas chromatography with flame photometric detection.

2.4.2 Silver foil

Non-saline waters

Mercury. Kalb[74] decomposed organomercury compounds in natural water with nitric acid. Liberated mercury was collected on a silver foil preconcentrator. Upon heating to 350 ° C the foil released a concentrated pulse of mercury which was then determined by cold vapour atomic absorption spectrometry.

References

1. Matthews, K. M. (1983). *Analytical Letters, 16*, 633.
2. Hashimoto, I., Satal, K., and Aoyage, M. (1985). *Journal of Radioanalytical and Nuclear Chemistry Articles*, *92*, 407.
3. Michel, J., Moore, W. S., and King, P. T. (1981). *Analytical Chemistry*, *53*, 1885.
4. Elsinger, R. J., King, P. T., and Moore, W. S. (1982). *Analytica Chimica Acta*, *144*, 277.
5. Butts, J., Todd, J. F., Berch, I., Moore, W. S., and Moore, D. G. (1988). *Marine Chemistry*, *25*, 349.
6. Rama, S., Todd, J. F., Butts, J. L., and Moore, W. S. (1987). *Marine Chemistry*, *22*, 43.
7. Harvey, B. R. and Dutton, J. W. R. (1973). *Analytica Chimica Acta*, *67*, 377.
8. Bacon, M. P. and Anderson, R. F. (1981). In *Trace Metals in Sea Water, Proceedings of a NATO Advanced Research Institute on Trace Metals in Sea Water*, 30/3–3/4/81, Sicily, Italy, eds C. S. Wong, *et al.*, Plenum Press, New York, p. 368.
9. Moore, W. S. and Reid, D. F. (1973). *Journal of Geophysical Research*, *78*, 8880.
10. Moore, W. S. (1976). *Deep Sea Research*, *23*, 647.
11. Moore, W. S. and Cook, L. M. (1975). *Nature (London)*, 253.
12. US Environmental Protection Agency (1975). In *Radiochemical Methodology for Drinking Water Regulations* EPA 600/4–75–005.
13. Ku, T. L., Huh, C. A., and Chen, P. S. (1980). *Earth Planet Science Letters*, *49*, 293.
14. Chung, Y. (1980). *Earth Planet Science Letters*, *49*, 319.
15. Moore, W. S. (1982). *Eustuarine, Coastal Shelf Science*, *12*, 713.
16. Reid, D. F., Key, R. M., and Schink, D. R. (1979). *Earth Planet Science Letters*, *43*, 223.
17. Michel, J., Moore, W. S. and King, P. T. (1981). *Analytical Chemistry*, *53*, 1885.
18. US Enviromental Protection Agency (1976). In *National Interium Primary Drinking Water Regulations*, EPA-57019–79–003.
19. Najiri, Y., Kawai, T., Otsuki, A., and Fuwa, K. (1985). *Water Research*, *19*, 503.

20. Prange, H., Omatsu, M., and Muskimoto, S. (1984). *X-Ray Spectrometry*, *13*, 83.
21. Sturgeon, R. E., Willie, S. N., and Berman, S. S. (1985). *Analytical Chemistry*, *57*, 6.
22. Irth, H., De Jong, G. J., Brinkman, U. A. T., and Frei, R. W. (1987). *Analytical Chemistry*, *59*, 98.
23. Sturgeon, R. E., Willie, S. N., and Berman, S. S. (1985). *Analytical Chemistry*, *57*, 2311.
24. Sunda, W. G. and Hanson, A. K. (1987). *Limnology and Oceanography*, *32*, 537.
25. Lieser, K. H., Quandt, S., and Gleitsmann, B. (1979). *Fresenius Zeitschrift fur Analytische Chemie, 298*, 378.
26. Zhang, Y., Riby, P., Cox, A. G., McLeod, C. W., Date, A. R., and Cheung, Y. Y. (1988). *Analyst (London)*, *113*, 125.
27. Schell, W. R., Nevissi, A., and Huntamer, D. (1978). *Marine Chemistry*, *6*, 143.
28. Bhatki, K. S. and Dingle, A. N. (1970). *Radiochemical and Radioanalytical Letters*, *3*, 71.
29. Fishman, M. J. (1970). *Analytical Chemistry*, *42*, 1462.
30. Muscat, V. I., Vickers, T. J., and Audrery, A. (1972). *Analytical Chemistry*, *44*, 218.
31. Matsunaga, K. (1974). *Mizushori-gijutsu*, *15*, 431.
32. Olafsson, J. (1974). *Analytica Chimica Acta*, *68*, 207.
33. Neske, P., Hellwig, A., Dornheim, L., and Triene, B. (1984). *Zeitschrift fur Analytische Chemie*, *318*, 498.
34. Alder, J. F. and Hickmann, D. H. (1977). *Analytical Chemistry*, *49*, 338.
35. Siemer, D. L. and Hageman, L. (1980). *Analytical Chemistry*, *52*, 105.
36. Chau, Y. K. and Saitoh, H. (1970). *Environmental Science and Technology*, *4*, 839.
37. Wittman, Z. (1981). *Talanta*, *28*, 271.
38. Temmerman, E., Dumarey, R., and Daws, R. (1985). *Analytical Letters (London)*, *18*, 203.
39. Kunert, I., Komarek, J., and Summiri, J. (1979). *Analytical Chimica Acta*, *106*, 285.
40. Doherty, P. E. and Dorsett, R. S. (1971). *Analytical Chemistry*, *43*, 1887.
41. Kaiser, I., Schock, P., and Togl, G. (1975). *Talanta*, *23*, 889.
42. Kalb, G. W. (1970). *Atomic Absorption Newsletter*, *9*, 54.
43. Fishman, M. J. (1970). *Analytical Chemistry*, *42*, 1462.
44. Muscab, U. I., Vickers, T. J., and Andrew, A. (1972). *Analytical Chemistry*, *44*, 218.
45. Olafsson, J. (1981). In *Trace Metals in Seawater, Proceedings of a Nato Advanced Research Institute on Trace Metals in Seawater*, 30/3–3/4/81, Sicily, Italy, eds C.S. Wong, *et al.*, Plenum Press, New York, p. 476.
46. Olafsson, J. (1974). *Analytica Chimica Acta*, *68*, 207.
47. Olafsson, J. (1982). *Marine Chemistry*, *11*, 129.
48. Gill, G. A. and Fitzgerald, W. F. (1987). *Marine Chemistry*, *20*, 227.
49. Syty, A., Christenson, R. G., and Rains, T. C. (1986). *Atomic Spectroscopy*, *7*, 89.

50. Cox, A. G., Cook, I. G., and McLeod, C. W. (1985). *Analyst (London)*, *110*, 331.
51. Desalvo, D. P. and Street, K. W. (1986). *Analyst (London)*, *111*, 1307.
52. Bagnati, R., Benfenati, E., Davoli, E., and Fanelli, R. (1988). *Chemosphere*, *17*, 59.
53. Sherma, J. (1988). *Journal of Liquid Chromatography*, *11*, 2121.
54. Thome, J. P. and Vandaele, Y. (1987). *International Journal of Environmental Analytical Chemistry*, *29*, 95.
55. Nyagah, G. (1981). *Journal of Chromatographic Science*, *19*, 500.
56. Fisher, R. (1982). *Fresenius Zeitschrift fur Analytische Chemie*, *311*, 109.
57. Sherma, J. (1986). *Journal of Liquid Chromatography*, *9*, 3433.
58. West, S. D. and Turner, L. G. (1988). *Journal of Association of Official Analytical Chemists*, *71*, 1049.
59. Ogan, K., Katz, E., and Slavin, M. (1978). *Journal of Chromatographic Science*, *16*, 517.
60. Nishikawa, Y. (1987). *Journal of Chromatography*, *392*, 349.
61. Sauer, W. A., Jadamse, J. R., Sager, R. W., and Killeen, T. J. (1979). *Analytical Chemistry*, *51*, 2180.
62. Nielen, M. W. F., Brinkman, U. A. T., and Frei, R. W. (1985). *Analytical Chemistry*, *57*, 806.
63. Maskarinec, M. P., Manning, D. L., Harvey, R. W., Griest, W. H., and Tomkins, B. A. (1984). *Journal of Chromatography*, *302*, 51.
64. Ghaoui, L. (1987). *Journal of Chromatography*, *399*, 69.
65. Noroozian, E., Maris, F. A., Nielen, M. W. F., Frei, R. W., De Jong, G. J., and Brinkman, U. A. T. (1987). *Journal of High Resolution Chromatography and Chromatography Communications*, *10*, 17.
66. Sackmauer, O. M., Pal'usova, O., and Szokolay, A. (1977). *Water Research*, *11*, 551.
67. Aue, W. A., Kapila, S., and Hastings, C. R. (1972). *Journal of Chromatography*, *73*, 99.
68. Hicks, E. and Riley, J. P., (1980). *Analytica Chimica Acta*, *116*, 137.
69. Lichiyama, M. (1978). *Water Research*, *12*, 299.
70. Semin, N. N., Filippov, Y. S., Toliskina, N. F., Smolyaninov, C. A., Volkov, S. A., and Kukushkin, V. S. (1986). *Journal of Chromatography*, *364*, 315.
71. Ogawa, I. and Fritz, J. S. (1985). *Journal of Chromatography*, *329*, 81.
72. Blaszkewicz, M., Baumhoer, G., and Neidhart, B. (1987). *International Journal of Environmental Analytical Chemistry*, *28*, 207.
73. Muller, M. D. (1987). *Analytical Chemistry*, *59*, 617.
74. Kalb, G. W. (1970). *Atomic Absorption Newsletter*, *9*, 84.
75. Syty, A., Christenson, R. G., and Rains, T. C. (1986). *Atomic Spectroscopy*, *7*, 89.

3

ACTIVE CARBON ADSORBENT

3.1 Metal cations

Two approaches have been used for the preconcentration of cations. Either the metals are chelated with an organic complexing agent and passed down a column of active carbon which adsorbs the metal complexes, or active carbon is modified by reaction with an organic chelating agent and then the solution of metal ions is passed through the column and thereby adsorbed. Examples of the former approach are the preconcentration of molybdenum as its ammonium pyrrolidinedithiocarbamate complex on charcoal[1] and preconcentration of manganese, iron, cobalt, nickel, copper, and zinc as their 8-quinolinates on activated carbon.[2] An example of the latter approach is the separation and preconcentration of traces of copper and lead from macro amounts of calcium magnesium, sodium, and potassium by adsorption from the sample on to active carbon modified with 8-hydroxyquinoline, dithizone or diethyldithiocarbamate.[3]

3.1.1 Non-saline waters

Atomic adsorption spectrometry Samchuk[4] developed an atomic absorption method for the determination of gold, silver, gallium, indium, thallium, cadmium, molybdenum, nickel, and copper in natural waters, with preconcentration on carbon modified by complex-forming organic reagents and on chelated sorbents. The methods for preparing the modified carbons and chelated sorbents are described. The work was done on a two-beam atomic absorption spectrometer with a graphite atomizer and deuterium background corrector. Hall *et al.*[5] preconcentrated gold from natural water on activated carbon discs prior to analysis by graphite furnace atomic absorption spectrometry. This enabled 10–15 ng litre^{-1} of gold to be determined.

Inductively coupled plasma atomic emission spectrometry Hall *et al.*[6] preconcentrated tungsten and molybdenum from natural spring water on to activated charcoal prior to analysis by inductivly coupled plasma atomic emission spectrometry and inductivly coupled plasma mass spectrometry. The detection limit for both elements by ICPMS was 0.06 μg litre^{-1}, and by ICPAES the detection limits were 1.2 and 0.4 μg litre^{-1}, respectively, for tungsten and molybdenum.

Neutron activation analysis Lieser *et al.*[7] applied neutron activation analysis to the determination of 15 trace elements with particular reference to the limits of detection obtained for different elements when employing a preconcentration technique based on adsorption of the metals as their dithizonates or diethyldithiocarbamates on charcoal. Metal analysis was carried out directly on the solid charcoal by neutron activation analysis. The method was applied to trace element determinations in river waters and North Sea waters. The 15 elements determined were cobalt, chromium, europium, iron, mercury, lanthanum, uranium, molybdenum, scandium, selenium, uranium, zinc, silver, gold, cadmium, and cerium for recoveries were between 75 and 100 per cent. For arsenic and antimony, recoveries were less satisfactory (55–70 per cent).

The following procedure was found to be most efficient for preconcentrating elements in estuary and sea water. One litre of water was first shaken at pH 5.5 with a suspension of 30 mg charcoal (Carbopurvor, 4 R, Degussa) and 1 ml of an acetone solution containing 1 mg sodium diethyldithiocarbamate. After filtration the solution is brought up to pH 8.5 by addition of 0.1 M ammonia and shaken again with a suspension of 30 ml charcoal and 1 ml of an acetone solution containing 1 mg dithizone. Both fractions of charcoal are combined, dried and filled into a quartz ampoule for irradiation.

X-Ray Fluorescence spectrometry Van der Bought and Van Grieken[2] optimized the conditions of the procedure for the preconcentration on carbon of some 20 elements including manganese iron, cobalt, nickel, copper, zinc, and cadmium as their 8-quinolinates (8-hydroxyquinolates). For a mixture of about 20 elements, they obtained an enrichment factor of 10000 a precision of 5 to 10 per cent, and an element recovery of 85–100 per cent depending on the element. Both X-ray fluorescence spectrometry and neutron activation analysis were used for the determination of preconcentrated metal 8-quinolinates in the activated carbon.

Van der Bought and Van Grieken[2] recommended the following optimal conditions for preconcentration of metals on activated carbon: The pH of the 1-litre water sample to be analysed is adjusted to a value of 8 using ammonium chloride/ammonia buffer. An optimal oxine quantity (see below) is added to provide a 5 mg litre^{-1} free oxine concentration. The oxine is dissolved by heating the sample to a temperature of 60 °C and the metal oxinate precipitate is filtered off after cooling. An optimal quantity of activated carbon dependent on the metal and oxine concentrations (see below) is added to the solution and an equilibration time of 1 h is taken into account. The activated carbon is filtered off and both loaded Nuclepore membrane filters are combined. For solutions of a definite type (e.g. potable, ground, surface, sea water, etc.) typical concentrations of trace elements can be supposed. These date together with the known calcium and magnesium concentration which can easily be determined by EDTA titration or flame emission spectrometry, allow the optimum oxine and activated carbon concentration to be calculated.

The oxine quantity necessary for leaving a 5 mg litre^{-1} free oxine concentration level after the complexation process can be calculated using the calcium and magnesium concentration and the 'mean ligand number' ($\bar{n}$) of the complexation reaction:

$$\bar{n} = \frac{C_L - [L]}{C_M} = \frac{\beta_1[L] + \ldots + N \cdot \beta_N[L]^N}{1 + \beta[L] + \ldots + \beta_N[L]^N} \quad (3.1)$$

where C_L and C_M are the analytical concentration of ligand and metal, respectively and [L] in the concentration of excess ligand in mol litre^{-1}. This quantity is given by

$$[L]\,(\text{mg Oxine litre}^{-1}) = \bar{n}\, C_M \frac{MW}{A} \quad (3.2)$$

where MW = molecular weight of the oxine, A = atomic mass of the metal and C_M = analytical concentration of metal ion, in mg litre^{-1}. A total oxine excess of 5 mg litre^{-1} corresponds to a free ligand concentration of 1×10^{-4} M (eqn 3.3) and mean ligand numbers $\bar{n}_{5Mg} = 0.66$ for magnesium ions and $\bar{n}_{5Ca} = 0.06$ for calcium ions, where the subscript points to a 5 mg litre^{-1} oxine excess. The total oxine quantity (Q) to be added per litre sample is given in eqn 3.4 when the molecular weight of oxine and atomic masses of magnesium and calcium are taken into account.

$$Q = S + E + 6\,\bar{n}_{5Mg} C_{Mg} + 3.6\,\bar{n}_{5Ca} C_{Ca} \quad (3.3)$$

where S is the estimated quantity necessary to complex the trace elements and E is the excess oxine, preferable around 5 mg litre^{-1}. Table 28, second column, lists S values for some typical water samples.

Some element recoveries obtained through the whole procedure including preconcentration and neutron activation analysis or X-ray fluorescence spectrometry are listed in Table 29. The minimum AC quantity necessary for quantitative adsorption of the metal oxinates can be calculated using a typical capacity for metal oxinate adsorption of 150 mg MOx mg^{-1} AC. The recommended AC quantities given in the third column of Table 28 include a five- to ten-fold excess over the minimum AC quantity to ensure quantitative recovery.

Van der Bought and Van Grieken[8] preconcentrated transition metals by conversion to oxinates and subsequent adsorption on to activated

Table 28 Oxine quantity S necessary to complex trace metals and AC quantity for quantitative enrichment of trace metals from different types water samples and approximate enrichment factor.

Sample	Estimated oxine quantity S[a] (mg litre^{-1})	AC quantity (mg litre^{-1})	Resulting enrichment factor
Potable water	5	100	10 000
Ground water	5	100	10 000
Surface water	15	150	6 666
Sea water	2	50	20 000

[a]See eqn (3.4).
From Van der Bought and Van Grieken[2] with permission.

Table 29 Per cent recovery for some elements.

Element	Recovery (per cent)	Element	Recovery (per cent)
Li	≈0	Se	≈0
Na	≈0	Ag	30
K	≈0	Cd	95
Cs	≈0	Sb	0–20
Mg	variable[a]	Hf	96
Ca	<10	Re	<99
Ba	≈0	Hg	90–100
Sc	56	Pb	80–90
Cr	90–100	Sm	91
Mn	90–100	Eu	86
Fe	90–100	Tb	89
Co	90–100	Dy	99
Ni	90–100	Yb	88
Cu	80–90	Lu	88
Zn	90–100		

[a]Enrichment by partial precipitation; recovery dependent on concentration.
From Van der Bought and Van Grieken[2] with permission.

carbon which was then examined by X-ray flourescence spectrometry or neutron activation analysis.

3.1.2 Sea-water

Neutron activation analysis Lieser *et al.*[9] studied the application of neutron activation analysis to the determination of trace elements in sea-water with particular reference to the limits of detection and reproducibility obtained for different elements when comparing different preliminary concentration techniques, such as adsorption on charcoal (also cellulose and quartz), and complexing agents such as dithizone and sodium diethyldithiocarbamate. In these procedures 1 litre of sea-water was shaken with 60 mg charcoal for 15 min. Complexing agents were added in amounts of 1 mg, dissolved in 1 ml acetone. The pH was 5.5 or it was adjusted to 8.5 by addition of 0.1 M ammonia. The charcoal was filtered off and irradiated. Results of three sets of experiments with charcoal alone, charcoal in the presence of dithizone, and charcoal in the presence of sodium diethyldithiocarbamate are presented in Table 30. The following elements are adsorbed to an extent from 75 to 100 per cent: silver, gold, cerium, cadmium, cobalt, chromium, europium, iron, mercury, lanthanum, scandium, uranium, and zinc. The amount of sodium is reduced to about 10^{-6}, bromine to about 10^{-5}, and calcium to about 10^{-2}. Analyses of North Sea water and suspended solids obtained by this procedure are tabulated in Table 31.

Van den Shoot *et al.*[10] preconcentrated molybdenum from sea-water as its pyrrolidinedithiocarbamate chelate on charcoal.

3.2 Anions

3.2.1 Non-saline water

Phosphate In a method developed to preconcentrate orthophosphate from natural water Hashitani *et al.*[11] used activated carbon loaded with zirconium. A 0.1–1.0 litre water sample at pH 1.5 was passed down a column of this material on which phosphate was adsorbed quantitatively and instantly below pH 8 and desorbed above pH 13.5. Pyrophosphate, tripolyphosphate, and metaphosphate behaved as orthophosphate.

Selenate and selenite Selenate and selenite can be determined by energy dispersive X-ray spectroscopy after preconcentration of elementary selenium on activated carbon, ascorbic acid being used to reduce selenium to its elemental form.[12–14]

3.3 Organics

Activated carbon filters have been employed for adsorption of different kinds of organics in natural waters since it was developed and introduced by the US Public Health Service.[15,16] However, the lack of adsorption and desorption control in addition to bacterial and oxidizing attack on the organics does place some limitations on the method.

Table 30 Mean values and standard deviations found by adsorption on charcoal (KF = correction factor).

Element	Concentration in the standardized water samples (g litre^{-1})	Found by adsorption on activated charcoal, pH 8.5; three determinations			Found by adsorption on activated charcoal in presence of dithizone pH 8.5; seven determinations			Found by adsorption on activated charcoal in presence of NaDDTC, pH 5.5; seven determinations		
		g litre^{-1}	per cent	KF	g litre^{-1}	per cent	KF	g litre^{-1}	per cent	KF
Ag	1×10^{-7}	$(8.5 \pm 1.4) \times 10^{-8}$	85	1.18	$(8.5 \pm 0.3) \times 10^{-8}$	85	1.18	$(5.5 \pm 1.5) \times 10^{-8}$	55	1.82
As	1×10^{-6}	$(6.6 \pm 1.2) \times 10^{-7}$	66	1.52	$(6.7 \pm 1.5) \times 10^{-7}$	67	1.49	$(3.2 \pm 1.9) \times 10^{-7}$	32	3.13
Au	1×10^{-9}	$(1.1 \pm 0.3) \times 10^{-9}$	100	1.00	$(1.1 \pm 0.1) \times 10^{-9}$	100	1.00	$(1.0 \pm 1.7) \times 10^{-9}$	100	1.00
Br	5×10^{-2}	$(5.3 \pm 1.0) \times 10^{-7}$	10^{-3}	–	$(2.1 \pm 0.5) \times 10^{-7}$	10^{-3}	–	$(3.6 \pm 1.7) \times 10^{-7}$	10^{-3}	–
Ca	1×10^{-1}	$(1.2 \pm 2.3) \times 10^{-3}$	1	–	$(1.7 \pm 0.2) \times 10^{-3}$	2	–	$(1.6 \pm 0.8) \times 10^{-3}$	2	–
Cd	1×10^{-4}	$(4.9 \pm 1.3) \times 10^{-7}$	49	2.04	$(9.5 \pm 2.7) \times 10^{-7}$	95	1.05	$(7.7 \pm 2.7) \times 10^{-7}$	77	1.30
Ce	1×10^{-6}	$(8.7 \pm 2.1) \times 10^{-7}$	87	1.15	$(8.2 \pm 1.1) \times 10^{-7}$	82	1.22	$(4.3 \pm 1.6) \times 10^{-7}$	43	2.33
Co	1×10^{-6}	$(4.0 \pm 0.6) \times 10^{-7}$	40	2.50	$(8.1 \pm 0.8) \times 10^{-7}$	81	1.23	$(7.6 \pm 1.0) \times 10^{-7}$	76	1.32
Cr	1×10^{-6}	$(9.1 \pm 0.3) \times 10^{-7}$	91	1.10	$(9.6 \pm 0.3) \times 10^{-7}$	96	1.04	$(3.6 \pm 0.2) \times 10^{-7}$	36	2.78
Eu	5×10^{-7}	$(5.2 \pm 0.4) \times 10^{-7}$	100	1.00	$(4.7 \pm 0.3) \times 10^{-7}$	95	1.05	$(3.8 \pm 1.7) \times 10^{-7}$	76	1.32
Fe	1×10^{-4}	$(7.4 \pm 1.5) \times 10^{-5}$	74	1.35	$(7.7 \pm 0.8) \times 10^{-5}$	77	1.30	$(7.0 \pm 0.6) \times 10^{-5}$	70	1.43
Hg	1×10^{-7}	$(9.7 \pm 0.2) \times 10^{-8}$	97	1.03	$(1.0 \pm 0.3) \times 10^{-7}$	100	1.00	$(1.0 \pm 0.1) \times 10^{-7}$	100	1.00
K	4×10^{-1}	$(3.2 \pm 0.4) \times 10^{-5}$	10^{-2}	–	$(2.1 \pm 0.3) \times 10^{-5}$	10^{-2}	–	$(4.3 \pm 0.9) \times 10^{-5}$	10^{-2}	–
La	1×10^{-6}	$(1.0 \pm 0.1) \times 10^{-6}$	100	1.00	$(1.0 \pm 0.1) \times 10^{-6}$	100	1.00	$(9.1 \pm 0.3) \times 10^{-7}$	91	1.10
Mo	1×10^{-6}	$(5.0 \pm 3.3) \times 10^{-7}$	50	2.00	$(2.1 \pm 0.9) \times 10^{-7}$	21	4.76	$(1.0 \pm 0.1) \times 10^{-6}$	100	1.00
Na	5	$(1.4 \pm 0.3) \times 10^{-5}$	10^{-4}	–	$(1.6 \pm 0.2) \times 10^{-5}$	10^{-4}	–	$(3.2 \pm 1.5) \times 10^{-5}$	10^{-4}	–
Sb	1×10^{-6}	$(1.8 \pm 0.5) \times 10^{-7}$	18	5.56	$(4.0 \pm 0.9) \times 10^{-7}$	40	2.50	$(5.6 \pm 2.0) \times 10^{-7}$	56	1.79
Sc	2×10^{-7}	$(1.9 \pm 0.2) \times 10^{-7}$	95	1.05	$(2.0 \pm 0.1) \times 10^{-7}$	100	1.00	$(1.4 \pm 0.1) \times 10^{-7}$	70	1.43
Se	1×10^{-4}	$(7.7 \pm 1.8) \times 10^{-7}$	77	1.30	$(6.1 \pm 1.7) \times 10^{-7}$	61	1.64	$(4.0 \pm 0.9) \times 10^{-7}$	40	2.50
U	1×10^{-7}	$(1.1 \pm 0.1) \times 10^{-7}$	100	1.00	$(1.2 \pm 0.2) \times 10^{-7}$	100	1.00	$(7.8 \pm 0.2) \times 10^{-8}$	78	1.28
Zn	1×10^{-6}	$(9.6 \pm 0.4) \times 10^{-7}$	96	1.04	$(1.0 \pm 0.1) \times 10^{-6}$	100	1.00	$(1.0 \pm 0.1) \times 10^{-6}$	100	1.00

From Lieser *et al.*[9] with permission.

Table 31 Determination of trace elements in sea-water (mean $\pm$ SD). Samples taken at 54°3′ North, latitude , 6°30′ east, longitude in the North Sea.

Element	Water without suspended material	Suspended material	Water with suspended material
Ag	$(8.8 \pm 0.4) \times 10^{-9}$	$(3.6 \pm 0.3) \times 10^{-9}$	$(8.7 \pm 0.4) \times 10^{-9}$
As	$(3.5 \pm 0.3) \times 10^{-7}$	$(3.4 \pm 1.2) \times 10^{-8}$	$(4.7 \pm 0.7) \times 10^{-7}$
Au	$(3.4 \pm 0.3) \times 10^{-10}$	$(4.5 \pm 2.0) \times 10^{-10}$	$(3.9 \pm 0.2) \times 10^{-10}$
Ba	$(5.7 \pm 0.1) \times 10^{-7}$	$(4.1 \pm 0.6) \times 10^{-7}$	$(1.3 \pm 0.5) \times 10^{-6}$
Br	$(5.5 \pm 0.1) \times 10^{-7}$	$(4.0 \pm 3.8) \times 10^{-6}$	$(3.7 \pm 0.2) \times 10^{-6}$
Ca	$(3.6 \pm 0.2) \times 10^{-5}$	$(1.5 \pm 0.3) \times 10^{-4}$	$(2.0 \pm 0.5) \times 10^{-4}$
Cd	$<10^{-6}$	$<10^{-6}$	$<10^{-6}$
Ce	$(3.4 \pm 0.3) \times 10^{-4}$	$(2.4 \pm 1.5) \times 10^{-6}$	$(1.0 \pm 0.2) \times 10^{-6}$
Co	$(4.5 \pm 0.3) \times 10^{-4}$	$(2.3 \pm 0.8) \times 10^{-4}$	$(6.7 \pm 0:2) \times 10^{-8}$
Cr	$(1.4 \pm 0.1) \times 10^{-7}$	$(1.3 \pm 0.3) \times 10^{-7}$	$(2.9 \pm 0.2) \times 10^{-7}$
Eu	$(8.2 \pm 1.4) \times 10^{-10}$	$(4.4 \pm 1.9) \times 10^{-9}$	$(5.6 \pm 1.7) \times 10^{-9}$
Fe	$(1.5 \pm 0.2) \times 10^{-5}$	$(3.5 \pm 1.2) \times 10^{-5}$	$(6.8 \pm 1.5) \times 10^{-5}$
Hg	$(2.2 \pm 0.2) \times 10^{-6}$	$<5 \times 10^{-9}$	$<5 \times 10^{-9}$
K	$(3.6 \pm 1.9) \times 10^{-5}$	$(2.9 \pm 0.2) \times 10^{-5}$	$(4.7 \pm 1.4) \times 10^{-5}$
La	$(3.2 \pm 0.1) \times 10^{-8}$	$(2.8 \pm 2.1) \times 10^{-8}$	$(7.4 \pm 1.4) \times 10^{-8}$
Mo	$(4.4 \pm 1.4) \times 10^{-8}$	$(2.1 \pm 0.1) \times 10^{-8}$	$(6.1 \pm 0.5) \times 10^{-8}$
Na	$(3.6 \pm 0.3) \times 10^{-4}$	$(4.7 \pm 6.9) \times 10^{-4}$	$(1.7 \pm 0.7) \times 10^{-4}$
Sb	$(5.7 \pm 0.4) \times 10^{-9}$	$(2.3 \pm 0.8) \times 10^{-9}$	$(1.3 \pm 0.2) \times 10^{-8}$
Sc	$(4.5 \pm 0.3) \times 10^{-8}$	$(1.9 \pm 2.0) \times 10^{-8}$	$(2.3 \pm 0.4) \times 10^{-8}$
Se	$(4.5 \pm 0.3) \times 10^{-8}$	$(3.5 \pm 0.9) \times 10^{-8}$	$(6.3 \pm 0.7) \times 10^{-8}$
U	$(3.3 \pm 1.0) \times 10^{-8}$	$(1.5 \pm 0.8) \times 10^{-8}$	$(5.6 \pm 0.9) \times 10^{-8}$
Zn	$(2.3 \pm 0.1) \times 10^{-6}$	$(3.3 \pm 1.5) \times 10^{-7}$	$(3.9 \pm 0.2) \times 10^{-6}$

From Lieser *et al.*[9] with permission.

3.3.1 Non-saline waters

Chlorinated insecticides Various workers[17-25] have investigated the use of activated carbon filters to concentrate chlorinated insecticides from water samples prior to gas chromatographic analysis. Quantitative values reported by these methods must be considered minimal because of lack of control on adsorption and desorption characteristics of the carbon filter bed and also because of bacterial breakdown of adsorbent organics during the sampling period. Using the carbon adsorption method determined that maximum organic adsorption efficiences are achieved at a reduced flow rate (not greater than 120 ml min^{-1} and total throughput volume of 1500 litres or less. In a study designed to determine carbon adsorption–desorption efficiency and reproducibility, it was determined by Eichelberger and Lichtenberg[27] that the carbon is useful for some organochlorines giving a 70–85 per cent recovery rate and less dependable for others, including DDT, which was determined to have an average 37 per cent recovery rate.

Eichelberger and Lichtenberg[28] used carbon adsorption prior to gas chromatography for the determination of methoxychlor, lindane, endrin, dieldrin, chlordane, DDT, heptachlor epoxide, and endosulfan in water samples. The eluate from the carbon was extracted with ethyl ether–hexane (3:17) and the carbon was dried and extracted with chloroform.

Brodtmann[29] carried out a long term study on the qualitative recovery efficiency of the carbon adsorption method versus that of a continuous liquid–liquid extraction method for several chlorinated insecticides. Comparative results obtained by electron capture gas chromatography indicate that the latter method may be more efficient. Samples of river water for analysis by the carbon adsorption method were collected at a rate of approximately 120 ml min^{-1} for a 7 day period in each case. At the end of the weekly sampling period, the total throughput volume was recorded. The carbon chloroform extract was then obtained by chloroform extraction of the carbon in a modified Soxhlet apparatus for 36 h using glass distilled, pesticide grade petroleum ether. The neutral fraction of the chloroform extract was then prepared for gas chromatography by the methods of Breidenbach.[30]

Breidenbach[30] used a continuous liquid–liquid extraction apparatus as described by Kahn and Wayman[31] for the extraction of non-polar solutes from river water. Pesticide grade petroleum ether, used in all cases, was recycled internally (initial solvent charge of 350 ml) thereby continuously exposing essentially fresh solvent to the river water. The throughput rate of river water was set at 33 ml min^{-1} for the sampling period of 7 days. To further enhance solute recoveries, three extractors were connected in series and solvent charges were pooled prior to concentration and clean up. The pooled extracts were then reduced in volume over a steam bath in flasks equipped with three-ball Snyder columns. A Florisil clean-up step using sequential elutions with 6 per cent and 15 per cent ethyl ether–petroleum ether solutions was employed (the Florisil was activated at 130 °C). A 300 mm × 120 mm plug of Florisil was topped by a 15 mm × 20 mm plug of anhydrous sodium sulphate and a wad of extracted glass wool. The Florisil was settled by tapping the tube lightly. This packing was then prewetted with 35 ml of hexane which was then discarded. When the last of the hexane reached the top of the Florisil packing, the sample was quantitatively transferred to the column. The sample was then eluted with 200 ml of 6 per cent v/v ethyl ether–petroleum ether solution. When the last of the 6 per cent eluate had reached the top of the packing, 200 ml of

15 per cent v/v ethyl ether–petroleum ether solution was added for the elution of dieldrin and endrin. Elution rates for both procedures were adjusted to approximately 5 ml min^{-1}.

At this stage, the 6 per cent eluate was ready for gas chromatography, but the 15 per cent eluate required a further clean-up step on a magnesium oxide–Celite 545 column. The magnesium oxide was prepared for use by forming a slurry of 200 g magnesium oxide in distilled water. The slurry was then heated on a steam bath for 30 min, vacuum filtered, and dried overnight at 130°C. A blender was used to pulverize the dried filtrate which was then mixed in a 1:1 ratio (w/w) with Celite 545. A second 300 mm × 20 mm i.d. chromatographic tube was attached to a 250 ml vacuum flask. A plug of glass wool was placed in the bottom of this tube, then 10 g of 1:1 magnesium oxide–Celite 545 was added under full vacuum (approximately 640 mm Hg) to pack it tightly. The vacuum line was then bled so that 35 ml petroleum ether was eluted at a rate of 15–20 ml min^{-1}. It was not necessary to discard this wash. After refluxed evaporation to 10 ml, the 15 per cent eluate was quantitatively transferred to the magnesium oxide–Celite 545 columns and eluted with 100 ml petroleum ether at the 15–20 ml min^{-1} rate with partial vacuum. This eluate was again evaporated to 15 ml by use of a Snyder column equipped flask and was then ready for gas chromatographic analysis as described below.

A dual column gas chromatograph equipped with two tritium electron capture detectors was empolyed by Brodtmann.[29] Both columns were acrylic glass, 1.83 mm long by 0.32 cm i.d. Column A was packed with 5 per cent DC-260 on 80/100 DCMS Chromosorb W. Flow rate of

Table 32 Overall recovery rates.

Insecticide	Percentage recovered	
	Carbon adsorption method	Liquid–liquid method
α-BHC	75.8	82.2
γ-Chlordane	78.4	95.2
p,p'-DDD	84.8	91.5
p,p'-DDE	55.6	93.8
o,p-DDT	74.0	95.9
p,p'-DDT	83.9	85.8
Dieldrin	85.2	95.6
Endrin	55.0	97.5
Heptachlor	52.6	83.8
Heptachlor epoxide	73.8	96.8
Lindane	77.5	92.2

carrier gas (5 per cent methane–argon) through this column was 80 ml min^{-1}. Column B was packed with 1.5 per cent OV-17/1.95 per cent QF-1 on 80/100 Chromosorb W DCMS support. Flow rate of carrier gas (5 per cent methane–argon) through this column was 50 ml min^{-1}. Both pairs of injectors, columns, and detectors were maintained at, respectively, 212, 184, and 204 °C.

Quantitation of unknown samples was accomplished by the comparison of peak heights for peaks in the unknown samples, confirmed by retention times on two different columns, to those of calibration mixtures of reference standard pesticides on both columns. Pesticide recoveries obtained by Brodtmann[29] are given in Table 32. These values were obtained by spiking carbon samples followed by extraction and clean-up for the carbon adsorption method and by spiking aliquots of petroleum ether after a 7 day sample period in the extractors. The overall conclusion of this work was that use of the liquid–liquid extractor method often finds low levels of chlorinated insecticides in water samples when the carbon adsorption method is insufficiently sensitive to do so.

Polychlorobiphenyls Berg *et al.*[32] separated polychlorobiphenyls from chlorinated insecticides on an activated carbon column prior to derivatization and gas chromatographic separation on the column. Separation is based on the observation that polychlorobiphenyls adsorbed on activated charcoal cannot be removed quantitatively with hot chloroform but can be with cold benzene. Insecticides of the DDT group and a variety of others (e.g. γ-BHC, aldrin, dieldrin, endrin, and heptachlor and its epoxide) can be eluted from the charcoal with acetone–ethyl ether (1:3). Typical recoveries from a mixture of *p,p'*-DDE (1,1-dichloro-2-bis-(4-chlorophenyl)ethylene), *p,p'*TDE, *o, p*-DDT, and a polychlorobiphenyl (Aroclor 1254) by successive elution with 90 ml of the 1:3 acetone–ether and 60 ml of benzene were 92, 92, 94, and 90 per cent respectively. Identification and determination of the polychlorobiphenyls was effected by catalytic dechlorination to bicyclohexyl or perchlorination to decachlorobiphenyl, followed by gas chromatography. For bicyclohexyl, gas chromatography was carried out on a column (8 ft × 0.25 in) in 10 per cent DC-710 on Chromosorb W, operated at 90 °C for 2.5 min then temperature programmed at 10 °C min^{-1} with flame ionization detection. For decachlorobiphenyl, the column (2 ft × 0.125 in) is 5 per cent SE-30 on Chromosorb W, operated at 215 °C, with nitrogen as carrier gas, and a tritium detector. Chriswell *et al.*[33] have also discussed the preconcentration of polychlorobiphenyls on charcoal.

Herbicides Di Corcia *et al.*[34] preconcentrated Simazine and Atrazine on a miniaturized graphitized carbon black column (Carbopack B). A 250 ml aqueous sample was passed through an adsorbent bed (50 mg) of graphitized carbon black (at an optimal flow rate of 30 ml min^{-1}. After washing with methanol (150 μl) the herbicides were desorbed with 0.7 ml of dichloromethane–methanol (3:2 v/v). Extracts were fractionated and analysed by reversed phase high performance liquid chromatography with a UV detector set at 220 nm. Using this rapid method recoveries of Simazine and Atrazine added to 250 ml of water at a concentration of 50 ng $litre^{-1}$ were 97.2 and 95.8 per cent with limits of detection of 0.07 ng and 0.15 ng, respectively.

Phenols Eichelberger *et al.*[35] collected the phenols from water samples on columns of activated carbon. The phenols were stripped from the carbon with chloroform prior to gas chromatography of the cleaned up extract. Clean-up was achieved by a treble extraction of the chloroform phase with aqueous sodium hydroxide. The aqueous extracts were combined, acidified with concentrated hydrochloric acid to pH 2, and extracted three times with ethyl ether, and the combined ether extracts are passed through a 10 cm Florisil column topped with anhydrous sodium sulphate (2 cm). The phenols are eluted with ether, and the eluate concentrated by evaporation and analysed by gas chromatography. An aluminium column (10 ft × 0.125 in o.d.) packed with 10 per cent of Carbowax 20 M terephthalic acid on HMDS-treated Chromosorb W (70–80 mesh) was used operated at 210 °C for the majority of phenols with nitrogen (50 ml min^{-1}) as carrier gas and flame ionization detection. Recoveries were on average, 85 per cent for alkylphenols and 94 per cent for chlorophenols.

Nitrosamines Fine *et al.*[36] and Kimoto *et al.*[37] have preconcentrated nitrosamines on carbon and subsequently desorbed them with chloroform, ethanol, or carbon disulphide prior to analysis by gas chromatography.

Miscellaneous organics Bacaloni *et al.*[38] listed adsorption data for 51 compounds, including alcohols, acids, phenols, ethers, aldehydes, ketones, esters, pesticides, and PCB, on graphitized carbon black at concentrations of 5–200 μg $litre^{-1}$ the compounds in water. The graphitized carbon black had a surface area of 100 m^2 g^{-1}, particle size 80–100 mesh, and pH 10.25. Chlorinated pesticides are among the compounds which are entirely adsorbed. Graphitized carbon black gave a larger specific retention volume and was more effective than Tenax, especially in non-potable water treatment. The best eluant for the recovery of chlorinated pesticides is 50:50 hexane: diethyl ether; recovery is generally 100 per cent with 5 ml of eluant.

Hertjies and Meijens[39] have conducted an examination of the effect of particle size, contact time, temperature, pH, and other factors on the adsorption of organics by active carbon. Mangani *et al.*[40] examined

Table 33 Use of active carbon in preconcentration of organics.

Substance preconcentrated	Method	Type of water	Ref.
Trihalomethanes	Minicolumn	Potable	42
Chlorinated hydrocarbons	–	Potable, waste water	43–45
Anionic detergents (linear alkylsulphonates)	–	–	46–49
Polyaromatic hydrocarbons	Short column	Potable	50
Metal chelates or aromatic and aliphatic phosphonates	Filtered through active carbon	Natural	51

the performance of graphitized carbon black (Carbopak B) in the preconcentration of some organic priority pollutants (pesticides, herbicides, phthalates, polyaromatic hydrocarbons) from natural water. In most cases 100 per cent recoveries were obtained in tests carried out by spiking either the adsorbent or the water sample. A very low eluent volume enabled high preconcentration ratios (1:1000). Coupled with gas chromatography–mass spectrometry, compounds off interest were determined at levels as low as 10 ppt.

Rivera *et al.*[41] desorbed organics from 2000-litre water supply samples on to granular activated carbon. Organics were then desorbed with dichloromethane, evaporated to dryness, dissolved in diethyl ether and fractionated. Soluble fractions were analysed by GC/MS/DS. Ether insoluble compounds were analysed and fractionated by HPLC with diode array detection followed by FAB, and FAB-CID-MIKE characterization. Two years' analytical data indicated that surfactants, plasticizers, ethylene glycol derivatives, phosphates, and hydrocarbons could be considered chromic pollutants in the rivers studied. The occurrence of 152 compounds was reported.

Other applications of carbon adsorption are summarized in Table 33.

3.3.2 Sea-water

Miscellaneous organics One of the earliest choices of adsorbent for sea-water organics was activated charcoal.[52,53] The technique has been refined for use in both fresh and sea-water by a number of workers.[54,55] A major problem in this technique has been the unknown efficiencies of collection and desorption. Jeffrey[56] using ^{14}C-labelled material, found that 80 per cent of the organic material in the sea-water was adsorbed by the charcoal. Of that 80 percent approximately 80 per cent again was desorbed by the solvents used. The overall efficiency of the method was therefore about 64 per cent. This method of collection is one of the few that permits the accumulation of gram amounts of organic materials from sea-water and therefore also permits the application of many of the standard techniques of organic analysis. However, we do not know how the distribution of compounds is changed in the process of adsorption and desorption; certain classes of compounds are probably entirely removed from the mixture, while others may be retained only in part. When so active a surface is used, there is also a possibility that chemical changes may take place during adsorption. The mixture as released from the charcoal may be considerably different from that originally present in sea-water. This collection method, although attractive for its simplicity and speed, is limited to qualitative results.

References

1. Sloot, H. A., Wals, G. D., and Das, H. A. (1977). *Analytica Chimica Acta*, *90*, 193.
2. Van der Bought, B. M. and Van Grieken, R. E. (1977). *Analytical Chemistry*, *49*, 311.
3. Zharikov, C. F. and Senyavin, M. M. (1970). *Trudy gas Okeanography Institute*, (1070) (101) Ref. *Zhur. Khim*. 19GD (7) Abstract No 7G189.
4. Samchuk, A. I. (1987). *Soviet Journal of Water Chemistry and Technology*, *9*, 57.
5. Hall, G. E. M., Vaive, J. E., and Ballantyne, S. (1986). *Journal of Geochemical Exploration*, *26*, 191.
6. Hall, G. E. M., Jefferson, C. W., and Michel, F. A. (1988). *Journal of Geochemical Exploration*, *30*, 63.
7. Lieser, K. H., Calmano, W., Heuss, E., and Neitzert, V. (1977). *Journal of Radioanalytical Chemistry*, *37*, 717.
8. Van der Bought, B. H., Van Grieken, R. E. (1980). *Talanta*, *27*, 417.
9. Lieser, K. H., Calmano, W., Heuss, E., and Neitzert, V. (1977). *Journal of Radioanalytical Chemistry*, *37*, 717.
10. Van der Shoot, H. A., Wals, G. D., and Das, H. A. (1977). *Analytica Chimica Acta*, *90*, 193.
11. Hashitani, H., Okumura, M., and Fujinaga, K. (1987). *Fresenius Zeitschrift fur Analytische Chemie*, *326*, 540.
12. Rubberecht, H. and Van Grieken, R. (1982). *Talanta*, *29*, 823.
13. Massee, R., Van der Sloot, H. A., and Das, H. A. (1977). *Journal of Radioanalytical Chemistry*, *38*, 157.
14. Orvini, E., Ladola, L., Gallorini, M., and Zerbia, T. (1981). In *Heavy Metals in the Environment*, Elsevien, Amsterdam, p. 657.
15. Braus, H., Middleton, F. M. and Walton, G. (1951). *Analytical Chemistry*, *23*, 1160.
16. Sproul, O. J. and Ryckman, D. W. (1961). *Journal of Water Pollution Control Federation*, *33*, 1188.

17. Goldberg, E. D. (ed.) (1972). In *A Guide to Marine Pollution*, Chapters 1, 2 and 4, Gordon and Breach, New York.
18. Braus, H., Middleton, F. M., and Walton, G. (1951). *Analytical Chemistry*, *23*, 1160.
19. Middleton, F. M., Grant, W., and Rossen, A. A. (1956). *Industry Engineering Chemistry*, *48*, 268.
20. Ludzok, F. J., Middleton, F. M., and Ettinger, E. B. (1958). *Sewage Industrial Wastes*, *30*, 622.
21. Polange, R. C. and Magrigian, S. (1958). *Journal American Water Works Association*, *50*, 1214.
22. Middleton, R. M. and Lichtenberg, J. J. (1960). *Journal Industrial Engineering Chemistry*, *52*, 99A.
23. Standard Methods for the examination of water and water APHA (13th edn), American Public Health Authority Washington, DC, p. 103 (1971).
24. Sproul, O. J. and Ryckman, D. W. (1961). *Journal Water Pollution Control Federation*, *33*, 1188.
25. Rosen, A. A. and Middleton, F. M. (1959). *Aalytical Chemistry*, *31*, 1729.
26. Booth, R. L., English, J. N., and McDermott, G. N. (1971). *Journal American Water Works Association*, *25*, 63.
27. Eichelberger, A. and Lichtenberg, J. J. (1971). *Journal of American Water Works Association*, *25*, 63.
28. Eichelberger, A. and Lichtenberg, J. J. (1971). *Journal of American Water Works Association*, *25*, 63.
29. Brodtmann, N. V. (1975). *Journal of American Water Works Association*, *67*, 558.
30. Breidenbach, A. W. (1968). In *Identification and Measurement of Chlorinated Hydrocarbon Pesticides in Surface Water*, 1968-0-315-842, US Govt Prntg Office, Washington, DC.
31. Kahn, L. and Wayman, C. H. (1964). *Analytical Chemistry*, *36*, 1340.
32. Berg, O. W., Diosady, P. L., and Rees, G. A. V. (1972). *Bulletin of Environmental Contamination and Toxicology*, *7*, 338.
33. Chriswell, C. D., Ericson, R. L., Junk, G. A., Lee, K. W., Fritz, J. S., and Svec, H. J. (1977). *Journal of American Water Works Association*, *69*, 669.
34. Di Corcia, A., Marchetti, M., and Samperi, R. (1987). *Journal of Chromatography*, *405*, 357.
35. Eichelberger, J. W., Dressman, R. C., and Longbottom, J. (1970). *Journal of Environmental Science and Technology*, *4*, 576.
36. Fine, D. H., Rounbehler, D. P., Huffman, F., and Epstein, S. S. (1975). *Bulletin of Environmental Contamination and Toxicology*, *14*, 404.
37. Kimoto, W. J., Dooley, C. J., Carre, J., and Fiddler, W. (1981). *Water Research*, *15*, 1099.
38. Bacaloni, A., Goretti, G., Lagana, A., Petronio, B. M., and Rotatori, M. (1980). *Analytical Chemistry*, *52*, 2033.
39. Hertjies, P. M. and Meijers, A. P. (1970). *Wasser Abwasser*, *111*, 61.
40. Mangani, F., Crescentini, P., Palma, P., and Bruner, F. (1988). *Journal of Chromatography*, *452*, 527.

41. Rivera, J., Ventura, F., Caixach, J., De Torres, M., and Figueras, A. (1987). *International Journal of Environmental Analytical Chemistry*, *29*, 15.
42. Dressman, R. C. and Stevens, A. A. (1983). *Journal of the American Water Works Association*, *75*, 431.
43. Hahn, J. (1982). *Korrespondenz Abwasser*, *29*, 539.
44. Heckel, E., (1986) *Umwelt*, *No. 1*, 31.
45. Kleopter, R. D. and Fairless, B. T. (1972). *Environmental Science and Technology*, *6*, 1036.
46. Wang, L. K., Kao, S. I., Wang, M. H., Kao, J. K., and Loshkin, A. L. (1978). *Industrial Engineering Chemistry Product Research and Development*, *17*, 186.
47. Wang, L. K., Wang, M. H., and Kao, J. T. (1978). *Water, Air and Soil Pollution*, *9*, 337.
48. Wang, L. K., Yang, J. Y., Ross, R. G., and Wang, M. H. (1975). *Water Research Bulletin*, *11*, 267.
49. Wang, L. K., Yang, J. Y., and Wang, M. H. (1973). In *Proceedings of 28th Industrial Waste Conference*, USA, 76–82.
50. Legana, A., Petronio, B. W., and Rotatori, M. (1980). *Journal of Chromatography*, *198*, 143.
51. Danz, H. O. and Jackwerth, E. (1984). *Fresenius Zeitschrift fur Analytische Chemie*, *318*, 22.
52. Jeffrey, L. M. and Hood, D. W. (1958). *Journal of Marine Research*, *17*, 247.
53. Wangersky, P. J. (1952). *Science*, *115*, 665.
54. Vaccaro, R. F. (1971). *Environmental Science and Technology*, *5*, 134.
55. Grob, K. and Zuercher, F. (1976). *Journal of Chromatography*, *117*, 285.
56. Jeffrey, L. M. (1969). In, *Development of a Method for Isolating Gram Quantities of Dissolved Organic Matter fron Sea Water and Some Chemical and Isotopic Characteristics of the Isolated Material*, PhD thesis, A & M. University, Texas, p. 152.

4

INORGANIC AND ORGANIC COPRECIPITATION TECHNIQUES

4.1 Inorganic coprecipitants – introduction

A further method of preconcentration involves adding to the sample a solution of a suitable metal such as iron, zirconium, or indium and following this by a precipitating agent which precipitates not only the added metal but also coprecipitates the metals which it is required to determine in the sample. The precipitate isolated from a large volume of original water sample is then isolated and dissolved in a small volume of acid to provide a concentrate for analysis.

4.2 Inorganic coprecipitants – Metal Cations

4.2.1 Ferric hydroxide

Non-saline waters

Spectrophotometric methods. Coprecipitation of arsenic with ferric hydroxide (added as ferric chloride) has been used for the preconcentration of mg litre^{-1} amounts of arsenic in potable water.[1,2] Pande and Sarin[2] have described a technique which can be used to enhance the sensitivity of the sodium diethyldithiocarbamate method for arsenic which involves coprecipitation of arsenic (arsenic(III) and arsenic(V)) with ferric hydroxide at pH 8.9 to 9.1 in ammonia medium, which preconcentrates the arsenic. The filtered ferric hydroxide is dissolved in a small volume of hydrochloric acid and the standard sodium diethyldithiocarbamate method is applied.

Reagents:

Ferric chloride (0.1 M): dissolve anhydrous ferric chloride (A. R. Grade) in distilled water. To prevent hydrolysis acidify with hydrochloric acid.

Standard ferric solution 1 g litre^{-1} iron.

Stock arsenic solution (1000 g litre^{-1} of arsenic(III) and arsenic(V) in hydrochloric acid (4 N). Store the stock solutions in polyethylene bottles.

Silver diethyldithiocarbamate solution: mix equal volumes of 0.1 M sodium diethyldithiocarbamate (2.25 g sodium diethyldithiocarbamate in 100 ml of distilled water) and 0.1 M silver nitrate (1.70 g silver nitrate in 100 ml distilled water) together in cold water. Allow the reaction mixture to stand for 5 min to complete the reaction. Precipitate the silver diethyldithiocarbamate out as yellow solid. Filter this precipitate through a sintered crucible, wash repeatedly with water, and finally dry over vacuum in a dessicator. Prepare a 0.5 per cent solution by dissolving 1 g of dried silver diethyldithiocarbamate in 200 ml of pyridine and store the solution in an amber coloured bottle.

Potassium iodide solution: 15 per cent solution (w/v) of potassium iodide in distilled water.

Lead acetate solution: 10 per cent lead acetate ($3H_2O$) w/v in distilled water. Use this solution to impregnate glass wool fitted in the scrubber.

Stannous chloride solution: 40 per cent solution (w/v) of stannous chloride $2H_2O$ in concentrated hydrochloric acid.

Zinc powder: arsenic free zinc powder of 20–30 mesh.

Hydrochloric acid: dilute 100 ml concentrated acid to 500 ml with distilled water.

Apparatus:

Spectrophotometer: Beckman Model DU Spectrophotometer or equivalent. Matched cells of 1.0 cm path length.

Arsine generation assembly, standard as described by Pande and Hasan.[3] The assembly consisted of:

(1) a 100 ml conical flask (Corning) with standard joints at the top and side delivery tube;

(2) a scrubber acting as a sulphide trap with standard joints at both ends;

(3) a delivery tube containing standard joints at one end and capillary of narrow bore at other end; and

(4) a narrow test tube.

Procedure:

(A) Preconcentration. Add 5.0 ml of 1000 mg litre^{-1} ferric chloride to 500 ml of water sample mix well. Adjust pH around 9.0. Keep this solution aside for about 2 h. Filter the solution through Whatman 42 paper. Dissolve the precipitate on filter paper in hydrochloric acid and wash it through the filter paper. Collect the washings carefully in a 50 ml volumetric flask and make up to volume.

(B) Arsenic estimation. Transfer 50 ml sample volume obtained after preconcentration in the arsine generator. Add 2 ml of potassium iodide and 1 ml of stannous chloride reagents and allow the mixture to stand for 20 min (this process facilitates the conversion of all the arsenic to trivalent state). Take 4.0 ml of silver diethyldithiocarbamate reagent in the absorbing tube. Add 3.0 g of zinc powder in the arsine generator with the help of a dry funnel and immediately close the stopper of the generator. Allow the liberated arsine to bubble through the reagent for 30 min. Transfer the absorbing solution to 1 cm cuvette and read the absorbance spectrophotometrically at 535 nm using silver diethyldithiocarbamate reagent as blank. Read the arsenic content from the calibration curve. The calibration curve can be prepared by taking absorbance values for standards ranging from 2 to 15 μg of arsenic in 50 ml solution.

Ferric hydroxide coprecipitation has also been used for the preconcentration of selenium in natural water.[4] Selenium (0.2–5 μg litre^{-1}) is coprecipitated with hydrated iron(III) oxide, dissolved in hydrochloric acid, reprecipitated as the element, and determined by molecular emission cavity analysis. For a 250 ml water sample, the detection limit is 0.2 μg litre^{-1}.

Reagents:

Sodium acetate, 2 M.

Ferric chloride, 0.1 M.

Hydrochloric acid, 8 M.

Hydroxylamine, 5 per cent aqueous.

Sulphur dioxide.

Apparatus:

Millipore filtration apparatus with glass fibre filter disc (2–3 mm diameter Whatman GF/C) supported on asbestos sheet disc.

Unicam SP90 atomic absorption spectrometer, modified to accommodate a sample holder assembly for the cavity. The cavity was the hexagonal aperture at the end of a 3 cm long mild steel screw into which the piece of filter paper with selenium deposit was placed. A Unicam emission burner head (2 cm o.d.) was used. The location of the cavity in the flame and flame conditions were set to give maximum intensity.[5]

Procedure:

Transfer a 100–250 ml water sample, containing 0–5 μg litre^{-1} selenium, to a 500 ml cylinder. Add 20 ml of 2 M sodium acetate solution followed by 1 ml of 0.1 M iron(III) chloride solution. After 12 h, filter the hydrated iron(III) oxide precipitate on a No. 4 Gooch crucible. Dissolve the precipitate in 10 ml of 8 M hydrochloric acid and add 0.5 ml of 5 per cent hydroxylamine solution, or bubble sulphur dioxide for 15 min. Heat at 70 °C for 3–5 min. Filter the hot suspension through a glass fibre filter disc (2–3 mm diameter, Whatman GF/C), which quantitatively retains the selenium, supported on an asbestos sheet disc in a Millipore filtration apparatus. Wash the selenium precipitate with a few ml of 8 M hydrochloric acid followed by hot water. Dry the filter disc and transfer it to the cavity so that it fits the contour of the cavity with the selenium deposit towards the aperture. Place the cavity in the flame and measure the maximal emission intensity at 411 nm. Determine the amount of selenium present by referring to a calibration graph prepared by measuring the emission of known amounts of selenium taken through the reduction and filtration procedure.

Preconcentration techniques coupled with a spectrophotometric finish have also been described for vanadium[6] and zirconium.[7]

X-Ray fluorescence spectrometry. Chakravatry and Van Grieken[8] used ferric hydroxide coprecipitation when determining traces of manganese, nickel, copper, zinc, and lead in natural waters including sea-water. Prior to analysis of the concentrate by energy dispersive X-ray fluorescence spectrometry, the optimum preconcentration procedure involved adding 2 mg iron to a 200 ml water sample, adding dilute sodium hydroxide up to pH 9, filtering off on a Nuclepore membrane after a 1 h equilibration time and analysing. Quantitative recoveries could then be obtained for nickel, copper, zinc, and lead at the 10 μ litre^{-1} level in waters of varying salinity while manganese was partially collected. The precision is 7–8 per cent at the 10 μg litre^{-1} level and the detection limits are in the 0.5–1 μg litre^{-1} range.

In Fig. 9 of Volume 1 is shown the effect of pH between 4 and 10 on the coprecipitation of 25 μg litre^{-1} manganese, nickel, copper, zinc, and lead with 10 mg litre^{-1} iron(III) as hydroxide. While traces of lead and zinc collect quantitatively beyond pH 6.0 and 8.0 respectively, copper and nickel exhibit a nearly quantitative coprecipitation only at pH 9.0 and above. In the case of manganese, the collection yield is only about 59–55 per cent at pH 9.0 and above. At the natural pH (8.0) of sea-water zinc and lead are coprecipitated totally while copper and nickel are collected up to about 80 per cent and manganese only about 11 per cent with an initial iron concentration of 10 mg litre^{-1}. As an optimum pH for the efficient collection of all the five trace elements, a pH of 9.0 was chosen.

Parallel precipitation experiments carried out on natural and synthetic sea-water, on 3 per cent sodium chloride solutions, and on bidistilled water samples, show at the strongly adsorbed ions, zinc and lead, are not affected by the changing matrix. Manganese and to a lesser extent copper and nickel appear to be collected significantly better in bidistilled water and sodium chloride solution than in sea-water. Probably the metal speciation is a most important factor in this context. However, Table 34 proves that no inhibitory effect due to the highly ionic complex

Table 34 Influence of sample matrix on the coprecipitation (2.5 μg metal spikes; 200 ml sample; pH 9; 1 h equilibration).

Element	Per cent coprecipitation⁻ ± standard deviation per measurement			
	Bidistilled water	3 per cent NaCl	North Sea water	Synthetic sea-water
Initial Fe-amount: 0.8 mg				
Mn	86 ± 9	93 ± 6	37 ± 6	
Ni	104 ± 7	100 ± 7	73 ± 4	
Cu	92 ± 4	95 ± 3	89 ± 9	
Zn	109 ± 7	106 ± 7	103 ± 13	
Pb	98 ± 5	91 ± 4	99 ± 10	
Initial Fe-amount: 2.5 mg				
Mn	98 ± 7	93 ± 3	48 ± 7	50 ± 10
Ni	112 ± 3	108 ± 8	95 ± 7	110 ± 3
Cu	101 ± 8	98 ± 4	99 ± 7	103 ± 2
Zn	104 ± 9	103 ± 4	106 ± 2	109 ± 4
Pb	98 ± 4	102 ± 4	99 ± 4	10 ± 4

From Chakravarty and van Grieken[8] with permission.

Table 35 Illustrative analyses of filtered water samples.

Element	Measured concentration (μg litre^{-1}) and stand. dev. per analysis				
	Estuary water	River water	Mine water[a]	Laboratory potable water	Bottled mineral drinking water
Mn	$\leqslant 0.4$	$\leqslant 0.6$	110 ± 20	$\leqslant 0.6$	$\leqslant 0.6$
Ni	2.0 ± 0.3	6.1 ± 0.2	50 ± 5	3.4 ± 0.2	$\leqslant 0.4$
Cu	2.3 ± 2.0	17 ± 5	$\leqslant 6$	47 ± 2	7.6 ± 1.9
Zn	11 ± 2	12 ± 1	8000 ± 600	70 ± 2	6.9 ± 2.9
Pb	1.6 ± 0.5	15 ± 10	62 ± 6	$\leqslant 0.8$	$\leqslant 0.8$

[a]Parallel NAA and SSMS on this sample yielded on the average: Mn: 120, Ni: 80, Cu: 2, Zn: 10900, Pb, 50 μg litre^{-1}.
From Chakravarty and Van Grieken[8] with permission.

matrix exists at low iron concentrations. A remarkably high effective preconcentration factor of 15 000 is achieved in this method.

To demonstrate the applicability of X-ray fluorescence analyses combined with iron hydroxide coprecipitation, Chakravarty and Van Grieken performed a number of analyses on estuary water, river water, potable water, bottled drinking water, and mine water as shown in Table 35. The results for the water canal sample and the underground mine water were derived from 200 ml samples.

Ferric hydroxide coprecipitation coupled with an X-ray fluorescence spectroscopic finish has also been used for the preconcentration and analyses of copper;[9] iron, zinc, and lead;[10] cadmium, lead, and copper;[11] and iron, zirconium, and tin[12] in natural waters. Others applications of ferric hydroxide coprecipitation are reviewed in Table 36.

Sea-water

Spectrophotometric methods. Manganese-54 has also been determined by a method[21] using coprecipitation with ferric hydroxide. The precipitate is boiled with hydrogen peroxide and the iron is removed by extraction with isobutyl methyl ketone. Zinc is separated on an anion exchange column and the manganese is separated by oxidizing it to permanganate in the presence of tetraphenylarsonium chloride and extracting the resulting complex with chloroform. Both ^{65}Zn and ^{54}Mn are counted with a 512-channel analyser with a well-type NaI(T1) crystal (3 × 3 in). Recoveries of known amounts of ^{65}Zn and ^{54}Mn were between 74 per cent and 84 per cent and between 69 per cent and 74 per cent respectively. Zirconium has been separated from seawater with 60–70 per cent efficiency by coprecipitation with ferric hydroxide prior to determination by Alizarin Red S spectrophotometric method.[41]

Table 36 Application of ferric hydroxide coprecipitation to the preconcentration of metals.

Metal	Type of water sample	Dissolving agents[a]	Analytical finish	Detection limit	Ref
Ga^{III}	Trade effluents	None	Scintillation spectrometry	–	13
Cr total and Cr^{III}	Natural	Hydrochloric acid	Atomic absorption spectrometry	0.2 μM	14
Cr^{III} Cr^{VI}	Natural	Hydrochloric acid	Atomic absorption spectrometry	Cr 40 ng $litre^{-1}$	15
Be, Cu, Zn, Pb, Bi, Co, Ni, Cd	Natural	Hydrochloric acid	Emission spectrography	1–10 μg $litre^{-1}$	16
Se^{IV}	Natural	–	Molecular emission cavity analysis	0.2 μg $litre^{-1}$	17, 18
Zr, Hf	Natural	–	Isotope dilution mass spectrometry	–	19
Pb, Cu, Zn, Cd	Natural	–	Differential pulse polarography	–	20

[a] If no dissolving agent for the ferric hydroxides is mentioned then the sample is analysed directly in the solid state.

Weisel *et al.*[42] determined aluminium, lead, and vanadium in Atlantic sea-water after coprecipitation on to ferric hydroxide.

Fluorimetric method. Shigematsu *et al.* determined cerium fluorimetrically at the 1 μg $litre^{-1}$ level in sea-water. Quadravalent cerium is coprecipitated with ferric hydroxide and the precipitate is dissolved in hydrochloric acid and interfering ions are removed by extraction with isobutyl methyl ketone. The aqueous phase is evaporated almost to dryness with 70 per cent perchloric acid, then diluted with water and passed through a column of bis-(2-ethylhexyl) phosphate on poly(vinyl chloride), from which Ce^{IV} is eluted with 0.3 M perchloric acid. The eluate is evaporated, then made 7 M in perchloric acid and treated with Ti^{III}, and the resulting Ce^{III} is determined spectrofluorimetrically at 350 nm (excitation at 255 nm).

Elderfield and Greaves[23] have described a method for the mass spectrometric isotope dilution analysis of rare earth elements in sea-water. In this method the rare earth elements are concentrated from sea-water by coprecipitation with ferric hydroxide and separated from other elements and into groups for analysis by anion exchange[24-29] using mixed solvents. Results for synthetic mixtures and standards show that the method is accurate and precise to ±1 per cent; and blanks are low (e.g. 10^{-12} moles La and 10^{-14} moles Eu). The method has been applied to the determination of nine rare earth elements in a variety of oceanographic samples.

Atomic absorption spectrometry. Cranston and Murray[30,31] took the samples in polyethylene bottles that had been precleaned at 20 °C for 4 days with 1 per cent distilled hydrochloric acid. Total chromium, $Cr^{IV} + Cr^{III} + Cr_p$ (particulate chromium), was coprecipitated with iron(II) hydroxide and reduced chromium ($Cr^{III} + Cr_p$) was coprecipitated with iron(III) hydroxide. These coprecipitation steps were completed within minutes of sample collection to minimize storage problems. The iron hydroxide precipitates were filtered through 0.4 μm Nuclepore filters and stored in polyethylene vials for later analyses in the laboratory. Particulate chromium, was also obtained by filtering unaltered samples through 0.4 μm filters. In the laboratory the iron hydroxide coprecipitates were dissolved in 6 M distilled hydrochloric acid and analysed by flameless atomic absorption. The limit of detection of this method is about 0.1–0.2 nmol $litre^{-1}$. Precision is about 5 per cent.

Nakayama *et al.*[32] have described a method for the determination of chromium (III), chromium-(VI), and organically bound chromium in sea-water. They found that sea-water in the sea of Japan contained about 9×10^{-9} M dissolved chromium. This is shown to be divided as about 15 per cent inorganic Cr^{III}, about 25 per cent inorganic Cr^{IV}, and about 60 per cent organically bound chromium. These workers studied the coprecipitation behaviours of chromium species with hydrated iron(III) and bismuth oxides.

The collection behaviour of chromium species was examined as follows. Sea-water (400 ml), spiked with 10^{-8} M Cr^{III}, Cr^{VI}, and Cr^{III} organic complexes labelled with ^{51}Cr, was adjusted to the desired pH by hydrochloric acid or sodium hydroxide. An appropriate amount of hydrated iron(III) or bismuth oxide was added; the oxide precipitates were prepared separately and washed thoroughly with distilled water before use.[33] After about 24 h, the samples were filtered on 0.4 μm Nuclepore filters. The separated precipitates were dissolved with hydrochloric acid and the solutions thus obtained were used for γ-activity measurements. In the examination of solvent extraction, chromium was measured by using ^{51}Cr, while iron and bismuth were measured by electrothermal atomic absorption spectrometry. The decomposition of organic complexes and other procedures were also examined by electrothermal atomic absorption spectrometry.

The percentage collection of Cr^{III} with hydrated iron(III) oxide may decrease considerably in the neutral pH range when organic materials capable of combining with Cr^{III}, such as citric acid and certain amino acids, are added to the sea-water.[34] Moreover, synthesized organic Cr^{III} complexes are scarcely collected with hydrated iron(III) oxide over a wide pH range.[34] As it was not known what kind of organic matter acts as the major ligand for chromium in sea-water Nakayama *et al.*[32] used EDTA and 8-quinolinol-5-sulphonic acid to examine the collection and decomposition of organic chromium species, because these ligands form quite stable water-soluble complexes with Cr^{III} although they are not actually present in sea-water. Both these Cr^{III} chelates are stable in sea-water at pH 8.1 and are hardly collected with either of the hydrated oxides. The organic chromium species were then decomposed to inorganic chromium(III) and chromium(VI) species by boiling with 1 g ammonium persulphate per 400 ml sea-water acidified to 0.1 M with hydrochloric acid. Iron and bismuth, which would interfere in atomic absorption spectrometry, were 99.9 per cent removed by extraction from 2 M hydrochloric acid solution with a *p*-xylene solution of 5 per cent tri-iso-octylamine. Cr^{III} remained almost quantitatively in the aqueous phase in the concentration range 10^{-9}–10^{-6} M, whether or not iron or bismuth was present. However, as about 95 per cent of Cr^{VI} was extracted by the same method, samples which may contain Cr^{VI} should be treated with ascorbic acid before extraction so as to reduce Cr^{VI} to Cr^{III}.

When the residue obtained by the evaporation of the aqueous phase after the extraction was dissolved in 0.1 M nitric acid and the resulting solution was used for electrothermal atomic absorption spectrometry, a negative interference, which was seemingly due to residual organic matter, was observed. This interference was successfully removed by digesting the residue on a hot plate with 1 ml of concentrated hydrochloric acid and 3 ml of concentrated nitric acid. This process had the advantage that the interference of chloride in the atomic absorption spectroscopy was eliminated during the heating with nitric acid.

Mullins[35] has described a procedure for determining the concentrations of dissolved chromium species in sea-water. Chromium(III) and chromium(VI) separated by coprecipitation with hydrated iron(III) oxide and total dissolved chromium are detemined separately by conversion to chromium(VI), extraction with ammonium pyrrolidinediethyldithiocarbamate into methyl isobutyl ketone and determination by atomic absorption spectroscopy. The detection limit is 40 ng litre^{-1} Cr. The dissolved chromium not amenable to separation and direct extraction is calculated by difference. In the waters investigated, total concentrations were relatively high (1–5 μg $litre^{-1}$) with chromium(VI) the predominant species in all areas sampled with one exception, where organically bound chromium was the major species. A standard contact time of 4 h was found to be necessary for the quantitative coprecipitation of chromium on ferric oxide. The r.s.d. values for the determination of chromium(III), chromium(VI), and total dissolved chromium were generally 10.0, 5.0, and 5.0 per cent, respectively. From these results, the r.s.d. for the calculated concentration of the bound species was 20 per cent.

Neutron activation analysis. The neutron activation method for the determination of arsenic and antimony in sea-water has been described by Ryabinin *et al.*[36] After coprecipitation of arsenic acid and antimony in a 100 ml sample of water by addition of a solution of ferric iron (10 mg iron per litre) followed by aqueous ammonia to give a pH of 8.4 the precipitate is filtered off and, together with the filter paper, is wrapped in polyethylene and aluminium foil. It is then irradiated in a silica ampoule in a neutron flux of 1.8×10^{13} neutrons cm^{-2} for 1–2 h. Two days after irradiation, the γ-ray activity at 0.56 MeV is measured with use of a NaI(Tl) spectrometer coupled with a multichannel pulse-height analyser, and compared with that of standards.

Gas chromatography. Siu and Berman[37] determined selenium in sea-water by gas chromatography after coprecipitation with hydrous iron(III) oxide. The method used rapid coprecipitation of selenium with hydrous iron(III) oxide. Following a brief stirring and settling period the coprecipitate was filtered and dissolved in hydrochloric acid, derivatized to 5-nitropiazselenol and extracted into toluene. The selenium was determined by gas chromatography–electron capture detection. The detection limit was 1 pg injected or 5 ng selenium per litre of sea-water using a 200 ml sample. Precision was 6 per cent at 25 pg selenium per litre.

Radiochemical methods. Silant'ev *et al.*[38] have described a procedure for the determination of strontium-90 in small volumes of sea-water. This method is based on the determination of the daughter isotope yttrium-90. The sample is acidified with hydrochloric acid, heated, and, after addition of iron, interfering isotopes are separated by double coprecipitation with ferric hydroxide. The filtrate is acidified with hydrochloric acid, yttrium carrier added, the solution set aside for 14 days for ingrowth of yttrium-90, and $Y(OH)_3$ precipitated from the hot solution with carbon dioxide free aqueous ammonia. Then $Y(OH)_3$ is reprecipitated from a small volume in the presence of hold-back carrier for strontium, the precipitate dissolved in the minimum amount of nitric acid, the solution heated, and yttrium oxalate precipitated by adding precipitated oxalic acid solution. The precipitate is collected and ignited at 800–850 °C to Y_2O_3. The cooled residue is weighed to determine the chemical yield, then sealed in a polyethylene bag and the radioactivity of the saturated yttrium measured on a low-background β-spectrometer. If the short-lived nuclides ^{140}Ba and ^{140}La are thought to be present in the sea-water sample, lanthanum carrier is introduced after the first $Y(OH)_3$ separation, and the system is freed from ^{140}La by precipitation of the double sulphate of lanthanum and potassium from a solution saturated with potassium sulphate.

Wong[39] has described a method for the radiochemical determination of plutonium in sea-water. This procedure permits routine determinations of ^{239}Pu activities (d.p.m.) down to 0.004 d.p.m. per 100 litres of sea-water. The plutonium is separated from sea-water by coprecipitation with ferric hydroxide. After further treatment and purification by ion exchange, the plutonium is electro-deposited on to stainless-steel discs for counting and resolution of the activity by α -spectrometry. For 30 samples the average deviation was generally well within the 1 so counting error. For sea-water the average recovery was 52 ± 18 per cent. The most serious interference is from ^{228}Th, which is present in most samples and is also a decay produce of the ^{236}Pu tracer.

Ballestra *et al.*[40] described a radiochemical measurement for determination of technetium-99 in rain, rivers, and sea-water which involved reduction to technetium(IV), followed by iron hydroxide precipitation and oxidation to the heptavalent state. Technetium(VII) was extracted with xylene and electrodeposited in sodium hydroxide solution. The radiochemical yield was determined by gamma counting on an anti-coincidence shield GM-gas flow counter. The radiochemical yield of 50- to 150-litre water samples was 20–60 per cent.

4.2.2 Zirconium hydroxide

Non-saline waters

Atomic absorption spectrometry. Shrivastava and Tandon[43] preconcentrated lead in natural or polluted waters by coprecipitation with zirconium hydroxide. This is done by adding zirconyloxychloride to water samples that have been brought to pH 8.2–9.5 with ammonia. The precipitate is redissolved in 50 per cent v/v nitric acid and its absorbance is measured at 283.2 mm. The relative error of the method with made up sea-water samples was about 3 per cent at the 10 μg litre^{-1} level. With polluted water samples the method was reliable for lead determination in the range 26–600 μg litre^{-1}.

Reagents:

Distilled water, triple glass distilled.

Stock lead solution 1000 mg $litre^{-1}$; dissolve 1.6000 g $Pb(NO_3)_2$ in 5 ml of 5 M nitric acid and make up to 1 litre.

Zirconium solution, 2 mg ml^{-1}. Dissolve 3.53 g $ZrOCl_2 \cdot 8H_2O$ in 10 ml of 6 M hydrochloric acid and make up to 500 ml.

Ammonia solution (20 per cent v/v).

Apparatus:

Pye Unicam SP 2900 atomic absorption spectrophotometer equipped with a lead hollow cathode lamp or equivalent.

Contamination from glassware, stoppers, and pipette tips was prevented by washing, with nitric acid (5 per cent v/v); after three rinses with deionized water and three rinses with further purified water, the glassware was stored under dust free conditions.

Procedure:

Adjust the pH of the water sample (200–250 ml) to 8.2–9.5 by the addition of ammonia solution (20 per cent v/v) and then add 5 ml zirconyloxychloride solution (2 mg ml^{-1}). Boil the solution for about 2 min, and allow to stand (precipitation) for about 10 min. Siphon off the supernatant waters, transfer the precipitate to a centrifuge tube, centrifuge, and dissolve in 2–3 ml of nitric acid (50 per cent v/v). Transfer the solution to a 5, 10, or 25 ml measuring flask depending upon the lead level being determined. A final volume of 10 ml is generally satisfactory. Instrumental settings (atomic absorption) used were 283.3 nm wavelength, 0.4 nm slit width, and air–acetylene as fuel. After extraction by the recommended procedure subtract the total blank from sample concentration.

To check the efficiency of this method for the determination of lead in sea-water, Shrivastava and Tandon[43] added spiked known amounts of lead into the water. The results are summarized in Table 37. The excellent agreement between the amount of lead added and found establishes the reliability of the method. The technique has also been used for the determination of ng $litre^{-1}$ levels of cadmium in fresh water.[44] In this method cadmium in a 1 litre sample, is coprecipitated with zirconium hydroxide, the precipitate is separated by dissolved air-flotation then dissolved in a small volume of dilute hydrochloric acid prior to analysis by atomic absorption spectrometry.

To investigate the applicability of this method to the separation and determination of cadmium in natural fresh water, Nakashima and Yagi[44] checked recoveries of known amounts of cadmium added to

Table 37 Determination of lead in natural-water samples.

Lead added (μg)	Lead found (μg)	Relative error (per cent)
1.00	0.92	−8.0
1.50	1.52	+1.3
2.00	2.06	+3.0
2.50	2.45	−2.0
3.00	3.06	+2.0
3.50	3.40	−2.9
5.00	4.90	−2.0

[a] Volume of sample = 200 ml.
From Shrivastava and Tandon[43] with permission.

fresh-water samples. For this purpose 1-litre aliquots of clear uncontaminated potable and river waters were filtered through 0.4 μm Nuclepore filters after the addition of 3 ml of hydrochloric acid per litre of sample immediately after collection. This system could be successfully applied to the separation and determination of cadmium at nanogram levels in fresh waters. The cadmium concentrations in the potable and river water samples were low: 7.8 and 18.7 ng litre^{-1}, respectively.

Nakashima and Yagi[45] have applied a similar procedure to the determination of copper, nickel, and cobalt in river and potable waters. Recoveries ranged from 96 to 102 per cent for spiked samples containing 1–100 μg litre^{-1} metal and from 94–99 per cent for potable and river water samples.

Sea-water

Inductively coupled plasma atomic emission spectrometry. Akagi *et al.*[46] used zirconium coprecipitation for simultaneous multi-element determinations of trace metals in sea-water by inductively coupled plasma atomic emission spectrometry. The coprecipitation procedure, ageing and washing of coprecipitates, and optimal pH conditions are described, together with spectral interferences. Recoveries of most metals increased with increase in pH except for hexavalent chromium and hexavalent molybdenum. Improved detection limits for 17 metals are reported including aluminium, arsenic, cadmium, cobalt, chromium, copper, iron, lanthanum, manganese, molybdenum, nickel, lead, antimony, titanium, vanadium, yttrium, and zinc.

4.2.3 Indium hydroxide

Sea-water

Inductively coupled plasma atomic emission spectrometry. Hiraide *et al.*[47] developed a multi-element preconcentration technique for chromium(III), manganese(II), cobalt, nickel, copper(II), cadmium, and lead in artificial sea-water using coprecipitation and flotation with indium hydroxide followed by inductively coupled plasma atomic emission spectrometry. The metals are simultaneously coprecipitated with indium hydroxide adjusted to pH 9.5 with sodium hydroxide, ethanolic solutions of sodium oleate and dodecyl sulphate added, and then floated to the solution surface by a steam of nitrogen bubbles. Cadmium may be completely recovered without the coprecipitation of magnesium. Concentrations of heavy metals (chromium(III), manganese(II), cobalt, nickel, copper(II), cadmium, and lead) in 1200 ml of artificial sea-water were increased 240-fold, while those of sodium and potassium were reduced to 2–5 per cent and those of magnesium, calcium, and strontium to 50 per cent. Down to 1 μg litre^{-1} of the above-mentioned heavy metals can be determined by this procedure. However, it is emphasized that real sea-water samples were not included in this study.

4.2.4 Lead phosphate and lead sulphate

Non-saline waters

Neutron activation analysis. Zhang *et al.*[48] preconcentrated various chromium species (total Cr, Cr^{IV}, Cr^{III}) in estuarine and sea water samples by coprecipitation with lead sulphate or lead phosphate prior to determination by neutron activation analysis and gamma spectrometry. Lead phosphate will collect both trivalent and hexavalent chromium whilst lead sulphate collects hexavalent chromium only. The procedure had a detection limit of 0.1 μg litre^{-1} for chromium in sea-water when 800 ml samples were used.

The method for coprecipitation of chromium(VI) with lead sulphate is as follows: Sodium sulphate (8 ml, 0.2 M) is added to an 800 ml aliquot of water sample filtered through a 0.45 μm membrane filter. The pH of the solution is adjusted to 3 with dilute nitric acid and 8 ml of 0.2 M lead nitrate is added dropwise under stirring. The reaction mixture is stirred for about 5 min and allowed to stand for about 1 h. The precipitate is filtered by suction using a Nuclepore membrane filter (pore size 0.4 μm) and washed with distilled water. The wet filter is folded into a 2/5 dram polyethylene vial and dried under an infra-red lamp. Up to 100 μg of chromium(III) does not interfere with the collection of 10 μg of chromium(VI) at pH 3.

Coprecipitation of total chromium with lead phosphate is carried out by following method. An 800 ml water sample (filtered through a 0.45 μm membrane filter) is acidified with 2 ml of (1 + 1) nitric acid; 3 ml of 0.2 M lead nitrate is added followed by 0.2 ml of 1 M ammonium dihydrogenphosphate and four drops of methyl red indicator; and (1 + 1) ammonia liquor is added dropwise with rapid stirring until the colour changes. The reaction mixture is allowed to stand for a few hours before filtration and the procedure is completed as in chromium(VI) collection. Both chromium(III) and chromium(VI) were quantitatively coprecipitated at pH 6.

Recoveries of both chromium(III) and chromium(VI) were excellent for both sample types with lead phosphate. Chromium(VI) was quantitatively recovered by the lead sulphate procedure from potable water but its recovery from sea-water was incomplete (87 per cent) because of the considerable amount of competing species, especially sulphate, present. Under the conditions used, the detection limit based on 2 $(\text{background})^{1/2}$ was 0.08 μg chromium, i.e. 0.1 μg litre^{-1} chromium for 800 ml samples. Coprecipitation as the phosphate has also been used[49] to preconcentrate lead in potable water at levels down to 50 μg litre^{-1} prior to polarographic analysis.

4.2.5 Barium sulphate

Non-saline waters Radium-228 has been preconcentrated from natural waters by coprecipitation with barium and lead sulphate. Radium-226 is then estimated in the precipitate by beta counting[50] or by alpha counting.[51] Barium sulphate impregnated alumina has been used to preconcentrate radium isotopes prior to their determination by X-ray spectrometry.[52] Acena and Crespo[53] preconcentrated radium-226 by coprecipitation with barium sulphate. De Jong and Wiles[54] preconcentrated thorium-230 by coprecipitation with barium sulphate. Mikac and Branica[87] preconcentrated inorganic lead from solutions also containing alkyl-lead species by coprecipitation with barium sulphate following the addition of barium chloride and sodium sulphate. Alkyl-lead compounds did not interfere in this procedure.

4.2.6 Metal sulphides

Non-saline waters Bart and Von Gunten[55] preconcentrated 20 trace elements in river water by coprecipitation with iron sulphide and ferric hydroxide prior to analysis by neutron activation analysis coupled with

Ge(Li) gamma ray spectrometry. The method was able to ascertain that all 20 elements were present at below recommended limits. Lead sulphide has been used to preconcentrate gold prior to determination by X-ray spectrometry. A detection limit of 0.2 μg litre^{-1} was achievable.[56]

Sea-water Patin and Morozov[57] developed a procedure for simultaneous concentration of mercury, lead, and cadmium from sea-water by coprecipitation with copper sulphide. The isolation yield is 99 per cent for mercury and lead and 89 per cent for cadmium. Mercury was determined by flameless atomic absorption spectrophotometry, and lead and cadmium by flame atomic absorption spectrophotometry.

4.2.7 Magnesium hydroxide

Non-saline waters Coprecipitation on to magnesium hydroxide has been used to preconcentrate tin(II).[58] The precipitate is dissolved in hydrochloric acid and determined in amounts down to 20 μg litre^{-1} by anodic stripping voltammetry. Partritnyi *et al.* preconcentrated tin by coprecipitation with magnesium hydroxide in alkaline medium followed by anodic stripping voltammetry at a graphite electrode at -1.0 V. Detection limits of 0.02 μg litre^{-1} were achieved.

Sea-water Petit[59] has described a method for the determination of tellurium in sea-water at picomolar concentrations. Tellurium(VI) was reduced to tellurium(IV) by boiling in 3 M hydrochloric acid. After preconcentration by coprecipitation with magnesium hydroxide, tellurium was reduced to the hydride by sodium borohydrate at 300 °C for 120 s then 257 °C for 12 s. The hydride was then analysed by atomic absorption spectroscopy. Recovery was 90–95 per cent and the detection limit 0.5 pmol litre^{-1}.

4.2.8 Miscellaneous inorganic coprecipitating agents

Non-saline waters These are summarized in Table 38.

Sea-water Nozaki and Tsunogai[70] determined lead-210 and polonium-210 in sea-water. The lead-210 and polonium-210 in a 30–50 litre sample are coprecipitated with calcium carbonate together with lead and bismuth and are then separated from calcium by precipitation as hydroxides. The precipitate is dissolved in 0.5 M hydrochloric acid, and polonium-210 is deposited spontaneously from this solution on to a silver disc and is determined by α-spectrometry. Chemical yields of lead and bismuth are determined in a portion of the solution from which the polonium has been deposited; hydroxides of lead and other metals are precipitated from the remainder of this solution and after

Table 38 Miscellaneous inorganic coprecipitating agents in natural waters.

Element preconcentrated	Coprecipitating agent	Carrier	Comments	Finish technique	Ref.
^{137}Cs	Ferrocyanide	Caesium	–	β-Counting	60
^{90}Sr		Carbonate	–	β-Counting	60
^{214}Bi		Bismuth	–	Scintillation counting	61
Au	Tellurium		–	AAS	62
Pb	Manganese dioxide	–	Acid solution lead converted to plumbane	Hydride AAS	63
^{214}Pb	–	Lead	–	Scintillation counting	64
^{32}P	Molydophosphate	–	–	Radiochemical	65
^{90}Sr	Carbonate	–	–	–	66,67
As, Sb, Bi, Se, Te	Lanthanum hydroxide	–	–	AAS	68
^{60}Co	Cobaltic hydroxide	–	Converted to hexane amino cobalt(III) complex which is extracted into methyl isobutyl ketone	Liquid scintillation counting	69

a period exceeding 3 months, the polonium-210 produced by decay of lead-210 is determined as before. The activity of lead-210 is calculated from the activity of polonium-210. The method was used to determine the vertical distribution of lead-210 and polonium-210 activities in surface layers of the Pacific Ocean.

Tsunogai and Nozaki[71] analysed Pacific Oceans surface water by consecutive coprecipitations of polonium with calcium carbonate and bismuth oxychloride after addition of lead and bismuth carriers to acidified sea-water samples. After concentration, polonium was spontaneously deposited onto silver planchets. Quantitative recoveries of polonium were assumed at the extraction steps and plating step. Shannon *et al.*[72] who analysed surface water from the Atlantic Ocean near the tip of South Africa, extracted polonium from acidified samples as the ammonium pyrrolidinedithiocarbamate complex into methyl isobutyl ketone. They also autoplated polonium onto silver counting disks. An average efficiency of 92 per cent was assigned to their procedure after calibration with ^{210}Po–^{210}Pb tracer experiments.

Leung *et al.*[73] describe a method for the separation and determination of uranium in sea-water. Thoric hydroxide ($Th(OH)_4$) was used as a collector. The final uranium concentration was measured via the fluorescence (at 575 nm) of its Rhodamine B. complex. The detection limit was about 200 μg litre^{-1}.

Mason[74] has described a rapid method for the separation of caesium-137 from a large volume of sea-water. In this procedure the sample (50 litres) is adjusted to pH 1 with nitric acid and ammonium nitrate (100 g), and caesium chloride (30 mg) is added as carrier. A slurry is prepared of ammonium molybdophosphate (7.5 g) and Gooch-crucible asbestos (715 g) with 0.01 M ammonium nitrate and deposited by centrifugation on a filter paper fitted in the basket of a continuous-flow centrifuge. The sample is centrifuged at 600–3000 r.p.m. and the deposit washed on the filter with 1 M ammonium nitrate (60–70 ml) and 0.01 M nitric acid (30–40 ml). The caesium collected on the filter is then prepared for counting by the method of Morgan and Arkell.[75] With this method the caesium can be extracted in less than 1 h.

Holm *et al.*[76] preconcentrated and determined neptunium-237 from large volume samples of sea-water. The procedure was based on the concentration of actinides from 1800-litre sea-water samples by hydroxide precipitations. Neptunium was isolated by ion exchange, fluoride precipitation, and extraction with thenoyltrifluoroacetone. Neptunium-239 or neptunium-235 was used as a radiochemical yield determinant. Neptunium was electroplated on to stainless-steel discs before alpha-spectrophotometry for about 10 days. The procedure allowed the sequential separation of plutonium, americium, technetium, and caesium, as well as neptunium. Although the radiochemical yield for neptunium was only 20–50 per cent, the procedure had been applied with success on several samples contaminated with neptunium-237 at fallout, or close to fallout, levels.

4.3 Inorganic coprecipitants – anions

4.3.1 Ferric hydroxide

Non-saline waters

Phosphorus-containing anions and arsenate. Hori *et al.*[77] examined the adsorption behaviour of phosphorus-containing anions as a function of solution pH. Adsorbed phosphorus compounds were determined spectrophotometrically and per cent adsorption calculated from adsorbed and initial amounts. This was plotted against solution pH. Orthophosphate was adsorbed quantitatively at pH 4.0–8.0, triphosphate and pyrophosphate at pH 4.0–9.3, and monomethylphosphate, phospite, and alpha-and beta-glycerophosphates at pH 4.0–6.8. Dimethylphosphate and hypophosphite were only slightly adsorbed at all pH values examined. The adsorbed ions were desorbed readily into a small volume of alkaline solution.

A solution of iron containing ferric chloride and acidified with hydrochloric acid is treated with aqueous ammonia to precipitate ferric hydroxide coprecipitated with ferric phosphate produced by any phosphate ions in the sample.[81] The precipitate is filtered off and dissolved in dilute hydrochloric acid and ammonia and uranyl acetate added to produce uranyl phosphate which is estimated polarographically. To determine phosphate plus arsenate the ferric hydroxide precipitate, containing coprecipitated iron phosphate and arsenite is made acid with hydrochloric acid and potassium iodide added. The arsenic trichloride produced is extracted with carbon tetrachloride and this solution back-extracted into hydrochloric acid. Application of the ammonia ferric chloride precipitation technique to this gives a preconcentrate containing arsenic only.

Chromate and dichromate. The sorption of hexavalent chromium (chromate) on to hydrous ferric oxide can be explained in terms of ligand exchange, hydroxyl ions being released from the surface of the hydrated oxide particles. For magnetic particles, however, the sorption effect was explained by the reduction of traces of hexavalent chromium at the magnetite/water interface, as a result of which a very fine ferric hydroxide/chromium hydroxide coating was produced which exhibited further reduction of hexavalent chromium.

4.3.2 Barium sulphate

Non-saline waters

Chromate and dichromate. The preconcentration of chromium(IV) (chromate) in natural waters by coprecipitation with barium sulphate has been shown to be a selective and accurate procedure.[79] Analysis of the preconcentrate is carried out for determining down to 0.02 μg litre^{-1} chromium spectrophotometrically. The method is based on the similarity between the solubility products of barium sulphate and barium chromate. Interference from ferric iron, aluminium, and trivalent chromium was overcome by using salicylic acid as a masking agent. The method was used on river water samples and the results obtained were in good agreement with those obtained by an aluminium hydroxide coprecipitation method. In order to prevent contamination, the membrane filter, the sample bottle, and other vessels were carefully precleaned with concentrated hydrochloric acid and redistilled water. The coprecipitation step for chromium(VI) was carried out as soon as possible after sampling to minimize the sample storage problem.

4.4 Inorganic coprecipitants – organics

Ferric hydroxide

Sea-water Ferric hydroxide is the precipitant most commonly used for the collection of organics. It is formed by the *in situ* formation of hydrated ferric oxides, usually by the addition of ferrous iron, followed by potassium hydroxide. The technique was first used for the precipitation of organic matter from an aged algal culture.[80] They recovered 79–95 per cent of the ^{14}C-labelled material from such cultures. Williams and Zirino[81] measuring efficiencies of removal of dissolved organic carbon, found that such scavenging collected between 38 and 43 per cent of the organic carbon measurable by wet oxidation with persulphate. Chapman and Rae[82] examined the effect of this precipitation on specific compounds. They found coprecipitation to be more complete with copper hydroxides, but still far from satisfactory. Only certain compounds were removed effectively by this treatment and the efficiency of removal varied with the water type and the organic compound involved.

4.4.2 Zirconium hydroxide

Sea-water A limited amount of work has been carried out using zirconium phosphates, compounds with well-defined coagulation and adsorption properties. The efficiency of coprecipitation was about 70 per cent for free amino acids and albumin. These methods may prove useful in the qualitative analysis of organic compounds, once the selectivities of the precipitants are understood. The metallic oxides suffer from the disadvantage of producing a precipitate which is difficult to filter, while calcite and zirconium phosphates produce relatively well-mannered precipitates. Even when the efficiences of collection of various model compounds in sea-water are known, the immense variety of organic compounds in sea-water will keep this technique largely qualitative.

4.4.3 Calcite

Sea-water The adsorption of organic matter on any surface presented to sea-water has been well documented. Neihof and Loeb[83] have demonstrated this adsorbance by following the change in surface charge of newly immersed surfaces. There has even been an attempt to use this phenomenon as a means of measuring dissolved organic carbon. Chave[84] found an association between calcite and dissolved organic materials in sea-water, and Meyers and Quinn[85] tried to use the effect as a method for the collection of fatty acids. As a collection technique, adsorption on calcite has several advantages. The pH of the sample is not greatly altered by the addition of small amounts of calcite; the precipitate is dense and should settle quickly; and after filtration the inorganic support can be removed by acidification. Unfortunately, the recovery of added fatty acids was inefficient; of the order of 18 per cent. Meyers and Quinn[86] achieved a somewhat greater efficiency of collection of fatty acids with clays, but the insolubility of the clays nullified one of the advantages of this concentration technique.

4.5 Organic coprecipitants – introduction

In some methods of analysis, e.g. X-ray fluorescence spectrometry, preconcentrated samples are required in the solid not the aqueous phase. In such methods the metal ion or ions to be determined are reacted with a suitable organic chelating agent under appropriate conditions to produce an insoluble complex. The small amount of precipitate thereby obtained from a large volume of water sample might then be left to 'age' prior to a careful filtration under controlled conditions and dried. The solid thereby obtained is suitable for metals analysis by X-ray fluorescence spectrometry of the solid, or after dissolution in a suitable reagent by liquid methods such as atomic absorption spectrometry.

4.6 Organic coprecipitants – metal cations

4.6.1 Dithiocarbamic acid derivatives

Non-saline waters

Atomic absorption spectrometry. Tris-pyrrolidinedithiocarbamate–cobalt III has been used to coprecipitate lead(II) prior to its determination by atomic absorption spectrometry.[88]

X-ray fluorescence spectrometry. Vanadium(IV) and vanadium(V) have been determined by energy dispersive X-ray fluorescence spectrometry following preconcentration by precipitation as the diethyldithiocarbamate.[89] Pentavalent vanadium was precipitated at pH 1.8 and tetravalent vanadium was precipitated with the same reagent at pH 4. The precipitates were collected by vacuum filtration on a membrane filter.

Experimental conditions for the preconcentration by precipitation of various metal complexes from water are discussed below:[90]

1. Sodium diethyldithiocarbamate (NaDDC). Samples (100 ml) were adjusted to pH 4 ± 0.05 and buffered with 2 ml of 0.1 M potassium hydrogen phthalate solution (pH 4) then 5 ml of aqueous 0.1 per cent (w/v) solution of sodium diethyldithiocarbamate was added with stirring. After the precipitate had aged for 15 min, the solution was filtered under vacuum through a 25 mm 0.45 – μm pore membrane filter and the filter was air-dried and mounted between Mylar films.

2. 1-Pyrrolidinecarbodithioic acid, ammonium salt (APDC). This procedure was as for sodium diethyldithiocarbamate with the exceptions that 1 ml of aqueous 1 per cent (w/v) APDC solution was added and the precipitates were aged for 20 min before filtration.

3. Sodium dibenzyldithiocarbamate (DBDTC). This procedure was as for sodium diethyldithiocarbamate with the exceptions that 1 ml of methanolic 1 per cent (w/v) solution was used and precipitates were aged for a total of 30 min.

4. Complexation with 8-quinolinol (oxine) and adsorption of complexes on activated carbon. Samples (100 ml) were adjusted to pH 8 ± 0.1 and buffered with ammonium chloride/ammonia solution (10 ml, pH 8). An amount of oxine solution (8 mg ml^{-1} in acetone) was added followed by 100 mg of activated carbon. The resulting suspension was then rotated in a glass vessel on a pot mill for 1 h, after which time the activated carbon was collected by filtration of the solution through membrane filter. The filter with the collected activated carbon was mounted between Mylar films while still moist. The lower Mylar support was perforated and the sample dried under vacuum.

5. Precipitation with thionalide and poly(vinylpyrrolidone) (PVP). To a 100 ml sample were added 5 ml of an aqueous 0.1 per cent (w/v) solution of poly(vinylpyrrolidone) and 5 ml of a 0.1 per cent (w/v) solution of thionalide (mercaptoacetic acid naphthylamide) in glacial acetic acid. The pH was adjusted to 4 ± 0.05 and the solution was filtered through a membrance filter (47 mm diameter, 0.45 cm pore size) after 15 min stirring and 5 min ageing. Then 25-mm diameter sections of these filters were punched out, air-dried, and mounted between Mylar films for X-ray analysis.

Following these preconcentration procedures, Ellis and Leyden[90] determined elements by X-ray fluorescence spectrometry.

On the basis of an extensive study of these various preconcentration methods based on precipitation, Ellis and Leyden[90] found that methods based on precipitation by a dithiocarbamate were optimal. These workers have also reported on the effect of interfering species in their methods.[91] The agents most likely to cause interference in actual sample analysis were found to be chromium(III), manganese(II), iron(III), cobalt(II), copper(II), zinc(II), lead(II), cadmium(II), sodium chloride, and calcium. It was found that for dithiocarbamate preconcentration methods, over half of the interference problems that were encountered were spectral in nature and compensation can be made for these. On the other hand, the other preconcentration methods suffered from mainly chemical interferences, for which corrections cannot be made.

Leyden *et al.*[92] used the following preconcentration procedures. The diethyldithiocarbamates were formed by adding 5 ml of the 0.1 per cent solution of the diethyldithiocarbamate sodium salt to a sample that has been adjusted to pH 4 with 2 ml of phosphate buffer. The solution was filtered through a Gelman filter (Metricet, Ga-6, 0.45 μm) after standing for about 15 min. The ammonium pyrrolidinedithiocarbamates were prepared similarly, but they were stirred for 2–3 min before ageing at least 20 min. The formation of the oxinates was accomplished by adjusting the pH value of the water samples to 8 ± 0.1 with ammonium chloride–ammonia buffer and adding a predetermined amount of oxine solution according to the recommendations given by Van Grieken and co-workers,[93,94] which takes into account the amount of oxine needed to complex calcium and magnesium and trace ions plus an excess of 5 mg litre^{-1} of reagent.

Recoveries obtained by energy dispersive X-ray spectrometry for several elements are tabulated in Table 39. The unsatisfactory nature of these and other date presented by Leyden *et al.*[92] indicate that much more work was required in the development of precipitation methods. The more recent work of Ellis *et al.*[90,91] would appear to be the more reliable.

Table 39 Recovery at 10 $\mu g\ litre^{-1}$ relative to recovery at 200 μg litre (per cent).

	Mn	Fe	Co	Cu	Zn	As	Hg	Pb
Sodium diethyl dithiocarbamate	—	51 ± 12	67 ± 16	115 ± 22	30 ± 28	—	61 ± 12	21 ± 9
Ammonium Pyrrolidinedithiocarbamates	—	72 ± 41	82 ± 14	84 ± 20	—	81 ± 15	46 ± 12	—
Activated carbon and oxinates	94 ± 27	—	109 ± 28	—	—	—	123 ± 22	110 ± 18

From Leyden *et al.*[92] with permission.

Caravajal *et al.*[95] have described a method for the preconcentration and determination of uranium in environmental water samples based on the coprecipitation of dissolved uranium in natural waters at pH 4 using an iron dibenzyldithiocarbamate carrier complex. The precipitate is collected as a thin film and measured by wavelength dispersive X-ray fluorescence spectrometry. Ultraviolet irradiation was used prior to coprecipitation to alleviate the effect of filter-clogging colloids such as humic acids. The iron carrier significantly improved the filtration time, while irradiation improved both the filtration time and the recovery from solutions containing below 10 μg of uranium. The results are precise and accurate. A detection limit of 0.4 μg litre^{-1} was achieved for 500 ml water samples with a counting time of 300 s. Of the ions tested, magnesium, calcium, sodium, potassium, sulphate, and chloride, none were observed to interfere at concentrations of 100 mg litre^{-1} or less. The applicability of the method to environmental water samples was assessed by spiking actual water samples and determining the amount of uranium present. The results showed the method to be reliable and applicable to environmental water samples.

The applicability of the method to the determination of uranium in natural waters was examined by analysing 100-ml river water samples spiked with known amounts of uranium. The water samples were chosen on the basis that the amount of uranium present (in 500 ml) was below the detection limit. At levels of less than about 20 μg, treatment with UV irradiation is essential to obtain full recovery of uranium from the water matrix.

The use of iron dibenzyldithiocarbamate as a coprecipitant for uranium in natural waters appears to be quite promising. Both UV irradiation and the addition of a carrier ion are essential if the applicability is to be extended to water samples where the concentration of uranium present is low (0.1–10 μg litre^{-1}) to provide for reasonable filterability and recovery. Although the iron carrier when absent did not appear to influence the precision, it had a dramatic effect on the filtration time of the sample; in contrast, irradiation improved both the filtration time and the recovery from solutions containing below 10 μg of uranium. It is apparent from extrapolation data that the linearity of the method extends past 50 μg. Although this extension may exist, the need for preconcentration becomes less essential. Thus, the preconcentration of uranium from natural waters with an iron dibenzyldithiocarbamate carrier complex prior to determination by X-ray spectrometry is precise, applicable to river water samples and free from interferences from major ions normally present in natural water. Judging from the results obtained for spiked natural water samples, the method would appear to give accurate as well as precise results.

Sea-water

Atomic absorption spectrometry. Copper, nickel, and cadmium have been preconcentrated and determined at the ng litre^{-1} level by coprecipitation with cobalt pyrrolidinedithiocarbamate followed by dissolution of the precipitate in an organic solvent and analysis by graphite furnace atomic absorption spectrometry.[96] Excellent results for the distribution of nickel and cadmium in the ocean were obtained by this technique.

X-Ray fluorescence spectrometry. Stiller *et al.*[97] have described the determination of cobalt, copper, and mercury in the Dead Sea by neutron activation analysis followed by X-ray spectrometry and magnetic deflection of β-ray interference. The metals were coprecipitated with lead ammonium pyrrolidinedithiocarbamate and detected by X-ray spectrometry following neutron activation. Magnetic fields deflect the β-rays while the X-rays reach the silicon (lithium) detector unaltered. The detectors have low sensitivity to γ rays. The concentration of cobalt found by this method was 1.3 μg litre^{-1} about one-fifth of that measured previously, while that of copper, 2.0 μg litre^{-1} agreed with results of some previous workers. The concentration of mercury was 1.2 μg litre^{-1}.

4.6.2 8-Hydroxyquinoline

Non-saline waters This has been used as a coprecipitant in the determination of aluminium, vanadium, copper, molybdenum, zinc, and uranium in lake waters, underground waters, and sea-water.[98] The final isolated solid was examined by neutron activation analysis.

Akatsuka and Atsuya[99] preconcentrated copper and manganese from river and potable water using 8-hydroxyquinoline and magnesium as carrier. Copper and manganese ions were coprecipitated at pH 7.0–8.5 (dilute ammonia solution) and were placed after filtering and drying in a mini-cup to be inserted in a graphite furnace atomic absorption spectrometer. Detection limits for copper and manganese were 12 and 14 ng litre^{-1}, respectively, for 300 ml portions of tap and river waters.

Caramella Crespi *et al.*[100] preconcentrated uranium as its oxinate on a bed of phenolphthalein and subsequently analysed the bed material by neutron activation analysis. Less than 0.5 μg litre^{-1} could be determined by this procedure.

4.6.3 8-Mercaptoquinoline and bis(8-quinolyl) disulphide

Non-saline waters Micro amounts of iron, cobalt, nickel, manganese, copper, and zinc have been preconcentrated from 1 litre natural water samples by adjustment to pH 1 and addition of sodium 8-mercaptoquinolinate solution. The pH was adjusted, after 15 min, to between pH 4–6 using 10 per cent ammonia. After 30 min, 1 per cent hydrogen peroxide (2 ml) was added and the solution stirred for 30 min before filtering to collect the precipitated metal complexes.

4.6.4 1,10-phenanthroline/tetraphenylboron

Non-saline waters Dogan and Haerdi[101] determined tin in water samples by preconcentrating by coprecipitation with 1,10-phenanthroline and tetraphenylboron. The precipitate was dissolved in ethyl alcohol and analysed by graphite furnace atomic absorption spectrometry in amounts down to 1 ng.

Sea-water Dogan and Haerdi[101] and Bergerioux and Haerdi[102] determined total tin in sea-water by graphite furnace atomic absorption spectrometry. These workers added 0.1–1.0 ml 0.25 M 1,10-phenanthroline and 0.1–1.0 ml 0.2 M tetraphenyl boron (both the reagents were freshly prepared) to 50–1000 ml of water sample which had been previously filtered through a 0.45 μm millipore filter. The pH of this solution was adjusted to 5.0 before addition of coprecipitating reagents. The precipitate thus obtained was either filtered or centrifuged and dissolved in a 1–5 ml aliquot of ammoniacal alcohol (methanol, ethanol or isopropanol) solution, pH = 8–9 or in Lumatom^R. For large volumes of water, the dissolution of the coprecipitate must be carried out with Lumatom^R since a precipitate is formed due to other ions present with ammoniacal alcohol solution.

4.6.5 1-(2-pyridylazo)-2-naphthol

Non-saline waters Bem and Ryan[103] described a procedure for the determination of seven trace elements in 800 ml samples of natural waters by neutron activation analysis after preconcentration with 1-(2-pyridylazo)-2-naphthol. This scheme was based on double irradiation of samples; for 10 min and for 16 h. Detection limits (μg litre^{-1}) were: cobalt (0.04), cadmium (0.8), copper (0.3), chromium (0.2), manganese (0.0006), uranium (0.006), and zinc (0.3).

Sea-water Atsuya and Itah[104] showed that 1-(2-pyridylazo)-2-naphthol used in conjunction with dimethyl glyoxime and nickel as a carrier ion is a very useful method for coprecipitating cadmium, copper, manganese, lead, and zinc prior to analysis by graphite furnace atomic absorption spectrometry. Coprecipitation with this complex increased the analytical sensitivity of the conventional graphite furnace method by 1000-fold. Detection limits ranged from 6 μg litre^{-1} (cadmium) to 50 ng litre^{-1} (lead) in lake and sea-water.

4.6.6 Thionalide (2-mercapto-N-2-naphthylacetamide)

Non-saline waters This reagent has been used[105] to precipitate from water samples, traces of silver, arsenic, cadmium, cobalt, iron, mercury, molybdenum, antimony, scandium, selenium, tungsten, and zinc. The trace elements were concentrated by coprecipitation with thionalide at pH 9.1. Coprecipitation with thionalide allowed the concentration of both ions and colloids. Ray and Johnson[106] preconcentrated arsenic by coprecipitation with thionalide prior to analysis by neutron activation analysis.

4.6.7 Thionalide–polyvinylpyrrolidone

Non-saline waters This mixture has been used to preconcentrate by precipitation a range of elements in the presence of calcium and magnesium.[107] Final analysis was X-ray fluorescence spectrometry. Analysis of selenium in the μg litre^{-1} range in water samples were achieved by this method.

References

1. Nakashima, S. (1978). *Analyst (London)*, *103*, 1031.
2. Pande, S. P. and Sarin, R. (1980). *Indian Journal of Environmental Health*, *22*, 189.
3. Pande, S. P. and Hasan, M. Z. (1978). *Indian Journal of Environmental Health*, *20*, 121.
4. Kouimtzis, T. A., Safoniou, M. C., and Papadoyannis, T. T. (1981). *Analytica Chimica Acta*, *123*, 315.
5. Belcher, R., Kouimtzist, T., and Townshend, A. (1974). *Analytica Chimica Acta*, *68*, 297.
6. Shcherbinina, S. D. and Petrova, S. Yu. (1972). *Energetik* (8), 21 Ref: *Zhur. Analit. Khim*. 19GD (23) Abstract No. 23G135 (1972).
7. Sastry, V. N., Krishnamoorthy, T. M., and Sarma, T. P. (1969). *Current Science (Bombay)*, *38*, 279.
8. Chakravarty, R. and Van Grieken, R. (1982). *International Journal of Environmental Analytical Chemistry*, *11*, 67.
9. Naito, W., Yoneda, A., and Azumi, T. (1979). *Water Purification and Liquid Wastes Treatment*, *20*, 329.
10. Bruninx, E. and Meyl, E. V. (1975). *Analytica Chimica Acta*, *80*, 85.
11. Laxen, D. P. H. (1984/5). *Chemical Geology*, *47*, 321.

12. Naito, W., Takahata, N., Yaneda, A., and Azumi, T. (1979). *Water Purification and Liquid Wastes Treatment*, *20*, 529.
13. Rafaeloff, R. (1972). *Radiochemistry and Radioanalytical Letters*, *9*, 373.
14. Cranstien, R. E. and Murray, J. W. (1978). *Analytica Chimica Acta*, *99*, 275.
15. Mullins, R. L. (1984). *Analytica Chimica Acta*, *165*, 97.
16. Lebedinskaya, M. P. and Chuiko, V. T. (1975). *Zhur. Analit. Khim.*, *28*, 863.
17. Belcher, R., Kouimtzis, T. A., and Townshend, A. (1974). *Analytica Chimica Acta*, *68*, 297.
18. Belcher, R., Bogdanski, S. L., Heyden, E., and Townshend, A. (1980). *Analytica Chimica Acta*, *113*, 13.
19. Boswell, S. M. and Elderfield, H. (1988). *Marine Chemistry*, *25*, 197.
20. Frimmel, F. H. and Gewwicz (1987). *Journal of Science of the Total Environment*, *60*, 57.
21. Stah, S. M. and Rao, S. R. (1972). *Current Science (Bombay)*, *41*, 659.
22. Shigematsu, T., Nishikawa, Y., Hiraki, K., Goda, S., and Tsujimatsu, Y. (1971). *Japan Analyst*, *20*, 575.
23. Elderfield, H. and Greaves, H. J. (1981). Trace Metals in Seawater. In, *Proceedings of a Nato Advanced Research Institute on Trace Metals in Seawater*. 30/3–3/4/81, Sicily, Italy, eds L. S. Wong *et al.*, Plenum Press, New York, p. 427.
24. Hooker, P. J. (1974). BA Thesis, University of Oxford.
25. Hooker, P. J., O'Nions, P. K., and Pankhurst, R. J. (1975). *Chemical Geology*, *16*, 189.
26. Korkisch, J. and Arrhenius, G. (1964). *Analytical Chemistry*, *36*, 850.
27. Faris, J. P. and Warton, J. W. (1962). *Analytical Chemistry*, *34*, 1077.
28. Desai, H. P., Krishnamoorthy, I. R., and Sandar-Das, M. (1964). *Talanta*, *11*, 1249.
29. Brunfelt, A. O. and Steinnes, E. (1969). *Analyst (London)*, *94*, 979.
30. Cranston, R. E. and Murray, J. W. (1978). *Analytica Chimica Acta*, *99*, 275.
31. Cranston, R. E. and Murray, J. W. (1980). *Limnology and Oceanography*, *25*, 1104.
32. Nakayama, E., Kuwamoto, T., Takoro, H., and Fujinaka, F. (1981). *Analytica Chimica Acta*, *131*, 247.
33. Nakayama, E., Kuwamoto, T., Tokoro, H., and Fujinaka, T. (1981). *Analytica Chimica Acta*, *130*, 401.
34. Takayama, E., Kuwamoto, T., Tokoro, H., and Fujinaka, T. (1981). *Analytica Chimica Acta*, *130*, 289.
35. Mullins, T. L. (1984). *Analytica, Chimica Acta*, *165*, 97.
36. Ryabinin, A. I., Romonov, A. S., Khatawov, S. L., Kist, A. A., and Khamidova, R. (1972). *Zhur. Analit. Chim.* *27*, 94.
37. Siu, K. W. H. and Berman, S. S. (1984). *Analytical Chemistry*, *56*, 1086.
38. Silant'ev, A. N., Chumichev, U. B., and Vakulouski, S. M. (1970). *Trudy Inst, eksp. Met. glav. uprav. gidromet, Sluzhty Sov. Minist. SSSR*, 15, (2) Ref: *Zhur Khim*. 19GD (1) Abstr. No. 1 G209 (1971).
39. Wong, K. M. (1971). *Analytica Chimica Acta*, *56*, 355.

40. Ballestra, S., Barci, G., Holm, E., Lopez, J., and Gastand, J. (1987). *Journal of Radioanalytical and Nuclear Chemistry*, *115*, 51.
41. Sastry, V. N., Krishamoorthy, T. M., and Sarnia, T. P. (1969). *Current Science (Bombay)*, *38*, 279.
42. Weisel, C. P., Duce, R. A., and Fasching, J. L. (1984). *Analytical Chemistry*, *56*, 1056.
43. Shrivastava, A. C. and Tandon, S. G. (1982). *International Journal of Environmental Analysis*, *12*, 169.
44. Nakashima, S. and Yagi, M. (1963). *Analytica Chimica Acta*, *147*, 213.
45. Nakashima, S. and Yagi, M. (1984). *Analytical Letters*, *17*, 1693.
46. Akagi, T., Nojiri, Y., Matsui, M., and Haraguchi, H. (1985). *Applied Spectroscopy*, *39*, 662.
47. Hiraide, M., Ito, T., and Baba, M. (1980). *Analytical Chemistry*, *52*, 804.
48. Zhang, H. F., Holzbecher, J., and Ryan, D. E. (1983). *Analytica Chimica Acta*, *149*, 385.
49. Spasojevic, B. D. and Tovanovic, D. A., (1971). *Acta Pharm. Jugoslavia*, *21*, 103.
50. Johnson, J. C. (1971). In, *US Geological Survey Water Supply Paper 1696G*.
51. Moran, M. C., Garcia-Tenorio, R., Garcia-Mantano, E., Garcia-Leon, M., and Madunga, G. (1986). *International Journal of Applied Radiation and Isotopes*, *37*, 383.
52. Perkins, R. W. (1979). In, *Report of Atomic Energy Commission. US BNWL 1051* Part 2 *Radium and Radiobarium measurement in seawater and freshwater by sorption and direct multidimensional X-ray spectrometry*.
53. Acena, M. L. and Crespo, M. T. (1988). *Journal of Radioanalytical and Nuclear Chemistry*, *126*, 77.
54. De Jong, I. G. and Wiles, D. R. (1984). *Water, Air and Soil Pollution*, *23*, 197.
55. Bart, G. and Von Gunten, H. R. (1979). *International Journal of Environmental Analytical Chemistry*, *6*, 25.
56. Petit, L. (1985). *Revues. Internationale d'Oceanographic Medicale*, *19*, 79.
57. Patini, A. M. and Morozov, N. P. (1976). *Zhur. Analitiches koi. Khimii*, *31*, 282.
58. Partretnyi, V. P., Malyuta, V. F., and Chinko, V. T. (1973). *Zhur. Analit Khim.*, *28*, 1337.
59. Petit, L. (1985). *Revue Internationale d'Oceanographic Medicale*, *19*, 79/80.
60. Alekson'Yan, O. M. (1972). *Gidrokim Mater*, *53*, 163.
61. Michaelis, M. L. (1969). *Chemikerzeiture Chemical Apparatus*, *93*, 883.
62. McHugh, J. R. (1983). *Applied Spectroscopy*, *4*, 66.
63. Vijan, P. N., Sadana, R. S. (1980). *Talanta*, *27*, 321.
64. Michaelis, M. L. (1969). *Chemikerzuture Chemical Apparatus*, *93*, 883.
65. Palagyi, S. (1983). *International Journal of Applied Radiation and Isotopes*, *34*, 755.
66. Gusev, N. G., Ya, U., and Margulis, A. N. (1966). In, *Dosimetric and Radiometric Methods Atomizdat*, Moscow, p. 444.

67. Barelta, E. J. and Knowles, F. R. (1986). *Journal of the Association of Official Analytical Chemists*, *69*, 540.
68. Thompson, M., Pahlavonpur, B., Walton, S. J., and Kirkbright, G. F. (1978). *Analyst (London)*, *103*, 568.
69. Claassen, H. C. (1970). *Analytica Chimica Acta*, *52*, 229.
70. Nozaki, Y. and Tsunogai, M. I. (1973). *Analytica Chimica Acta*, *64*, 209.
71. Tsunogai, S. and Nozaki, Y. (1971). *Geochemical Journal*, *5*, 165.
72. Shannon, L. V., Cherry, R. D., and Orren, M. F. (1970). *Geochimica and Cosmochimica Acta*, *34*, 701.
73. Leung, G., Kim, Y. S., and Zeitlin, H. (1972). *Analytica Chimica Acta*, *60*, 229.
74. Mason, W. J. (1974). *Radiochemical and Radioanalytical Letters*, *16*, 237.
75. Morgan, A. and Arkell, O. (1963). *Health Physics*, *9*, 857.
76. Holm, E., Aarkrog, A., and Ballestra, S. (1987). *Journal of Radioanalytical and Nuclear Chemistry*, *115*, 5.
77. Hori, T., Moriguichi, M., Sassaki, Kitagawa, S., and Munakato, M. (1985). *Analytica Chimica Acta*, *173*, 299.
78. Rozaiski, L. (1972). *Chimica Analit*, *17*, 55.
79. Yamazaki, H. (1980). *Analytica Chimica Acta*, *113*, 131.
80. Jeffrey, L. M. and Hood, D. W. (1958). *Journal of Marine Research*, *17*, 247.
81. Williams, P. M. and Zirino, A. (1964). *Nature (London)*, *204*, 462.
82. Chapman, G. and Rae, A. C. (1967). *Nature (London)*, *214*, 627.
83. Neihof, R. and Loeb, G. (1974). *Journal of Marine Research*, *32*, 5.
84. Chave, K. E. (1965). *Science*, *148*, 1723.
85. Meyers, P. A. and Quinn, J. G. (1971). *Limnology and Oceanography*, *16*, 992.
86. Meyers, P. A. and Quinn, J. G. (1973). *Geochimica and Cosmochimica Acta*, *37*, 1745.
87. Mikac, N. and Branica, M. (1988). *Analytica Chimica Acta*, *212*, 349.
88. Krishnamurty, K. V. and Reddy, M. M. (1977). *Analytical Chemistry*, *49*, 222.
89. Hirayama, K. and Keyden, D. (1986). *Analytica Chimica Acta*, *188*, 1.
90. Ellis, A. T. and Leyden, D. E. (1982). *Analytica Chimica Acta*, *42*, 73.
91. Ellis, T. A., Leyden, D. E., Wegscheider, W., Jablonski, B. B., and Bodnar, W. B. (1962). *Analytica Chimica Acta*, *142*, 89.
92. Leyden, D. E., Wegscheider, W., and Bodnar, W. B. (1979). *International Journal of Environmental Analytical Chemistry*, *7*, 85.
93. Vanderborght, B. M., Verbeek, J., and Van Grieken, R. E. (1977). *Bulletin of the Belgian Chemical Society*, *86*, 23.
94. Vanderborght, B. M. and Van Grieken, R. E. (1977). *Analytical Chemistry*, *49*, 311.
95. Caravajal, G. S., Mahan, K. L., and Leyden, D. E. (1982). *Analytica Chimica Acta*, *135*, 205.
96. Boyle, I. A. and Edmond, J. M. (1977). *Analytica Chimica Acta*, *91*, 189.
97. Stillet, M., Mantel, M., and Rappaport, M. S. (1984). *Journal of Radioanalytical and Nuclear Chemistry*, *83*, 345.
98. Fujinaga, T., Kusaka, R., Koyama, M., Tsuji, H., Mitsujim, T., Imai, S.,

Okuda, J., Takamatsu, T., and Ozaki, T. (1973). *Journal of Radioanalytical Chemistry*, *13*, 301.
99. Akatsuka, K. and Atsuya, I. (1987). *Analytica Chimica Acta*, *202*, 223.
100. Caramella Crespi, V., Genova, N., Melonis, N., and Oddone, M. (1987). *Journal of Radioanalytical and Nuclear Chemistry*, *114*, 303.
101. Dogan, S. and Haerdi, W. (1980). *International Journal of Environmental Analytical Chemistry*, *8*, 249.
102. Bergerioux, C. and Haerdi, W. (1980). *Analysis*, *8*, 169.
103. Bem, H. and Ryan, D. E. (1984). *Analytica Chimica Acta*, *166*, 189.
104. Atsuya, I., and Itah, K. (1988). *Fresenius Zeitschrift fur Analytische Chemie*, *329*, 750.
105. Zmijewska, W., Polkowska Motrenko, H., and Stakowska, H. (1984). *Journal of Radioanalytical and Nuclear Chemistry Articles*, *84*, 319.
106. Ray, B. J. and Johnson, D. E. (1972). *Analytica Chimica Acta*, *62*, 196.
107. Panayappan, R., Venezky, D. L., Gilfrich, J. V., and Birks, L. S. (1978). *Analytical Chemistry*, *50*, 1125.

5

SOLVENT EXTRACTION OF ORGANIC AND ORGANOMETALLIC COMPOUNDS

5.1 Organic compounds – Non-saline waters

Solvent extraction is an attractive method for concentrating a particular fraction of the dissolved organic matter, the fraction concentrated being determined by the choice of solvent. The most obvious limitation on the method is that set by solvent choice, the solvent should have only limited solubility in water, which limits the materials removed to the less polar compounds. Another limitation is that set by contamination. The solvent used must be purified carefully, since the amounts of the various organic compounds collected from water could be about as large as the trace impurities in the solvents. Once the separation into the organic solvent has been accomplished, any of a number of techniques of fractionation and analysis can be applied. The first step in the solvent extraction is the actual sampling of the water. All of the problems associated with sampling can occur in this step and may be aggravated by the large volumes of water customarily employed.

Solvent extraction has been applied both as a batch and a continuous process[1-5] and to natural waters and sea-waters. Since a large mass of water is to be extracted, the batch methods are limited although Kawahara *et al.*[6] constructed a semiautomatic device to speed up the procedure and reduce the organic solvent volume. Batch methods are usually constructed without continuous refreshing systems. Instead a repeated shaking with a fresh portion of solvent is used.

Various workers[7-9] have described an *in situ* solvent extraction equipment suitable for on-site operation on river banks etc. Ahnoff and Josefsson[7] point out that to avoid sample loss and contamination it is desirable to carry out sample enrichment for the analysis of trace organic compounds in water as near to the sampling point as possible. Even sampling points on adjacent land stations or ships can lead to contamination from tubing walls and the inner surfaces of the pump and also deposition of particles can occur in the tubing. The *in-situ* apparatus they describe is designed to perform solvent extraction of large amounts of water, using the apparatus continuously while situated at the sampling point at a desired depth.

Peters[2] has compared continuous extractors for the extraction and concentration of trace organics from water. Baker *et al.*[3] evaluated a Teflon helix liquid–liquid extraction for concentration of trace organics in water into methylene chloride. Gomella *et al.*[4] devised a method of extracting and concentrating micropollutants in water with cyclohexane. The cyclohexane extracts are frozen and the cyclohexane is then allowed to sublime. Fritz[5] has compared various methods for concentrating trace organics from water including resin sorption, solvent extraction, water gas distribution, gas stripping, direct injection into a gas chromatograph and resin sorption with thermal desorption. Burchill *et al.*[10] give details of equipment and procedure for the routine monitoring of water for trace organic compounds down to the 0.01 μg litre^{-1} level involving solvent extraction with dichloromethane followed by gas chromatography with flame ionization detection and mass spectrometry.

A liquid–liquid extraction procedure for the simultaneous recovery of a wide range of basic, neutral, and phenolic organic compounds in a single container has been described.[11] Samples were extracted with a mixture of dichloromethane and ether, using a specially designed strirrer and modified micro Snyder evaporator. Analysis was possible by various chromatographic methods and recoveries varied between 72 and 109 per cent with a detection limit of 0.01 μg litre^{-1} for each compound.

Eichelberger *et al.*[12] studied the effect of sample pH on precision and accuracy in the determination of organics in water by methylene dichloride extraction followed by fused silica capillary column gas chromatography—mass spectrometry and packed column gas chromatography—mass spectrometry.

Murray[13] has described a rapid micro-extraction procedure for preconcentration and analysis of trace amounts of organic compounds in water and compared the results obtained with those obtained by macro-extraction methods. Final analysis of the solvent extract was carried out by gas chromatography. The method using a flask with a side arm and fitted with a capillary tube involved shaking 980 ml of water with 200 μl of hexane. By carefully adding water through the side arm the solvent layer was displaced into the capillary tube after tilting the flask. About 50 μl were recovered and were suitable for direct analysis by gas chromatography. This method gave as good a recovery and was much more rapid than macro-extraction or continuous extraction using three inverted flasks with a vibratory mixer and siphon system.

In the micro method an extraction flask (Fig. 4(a)) similar to the rapid liquid extraction used by Grob *et al.*[14] was used. The flask, which contained 980 ml of water and 200 μl of hexane, was manually shaken for 2 min. By tilting the extraction flask and carefully adding water through the side arm, the solvent layer could be held in the centre portion and finally displaced into the capillary tube. About 50 μl could be recovered and was suitable for direct analysis by gas chromatography.

For the macro method a steam distillation solvent extraction head similar in design to No. 6555, manufactured by Ace Glass Co. (as used

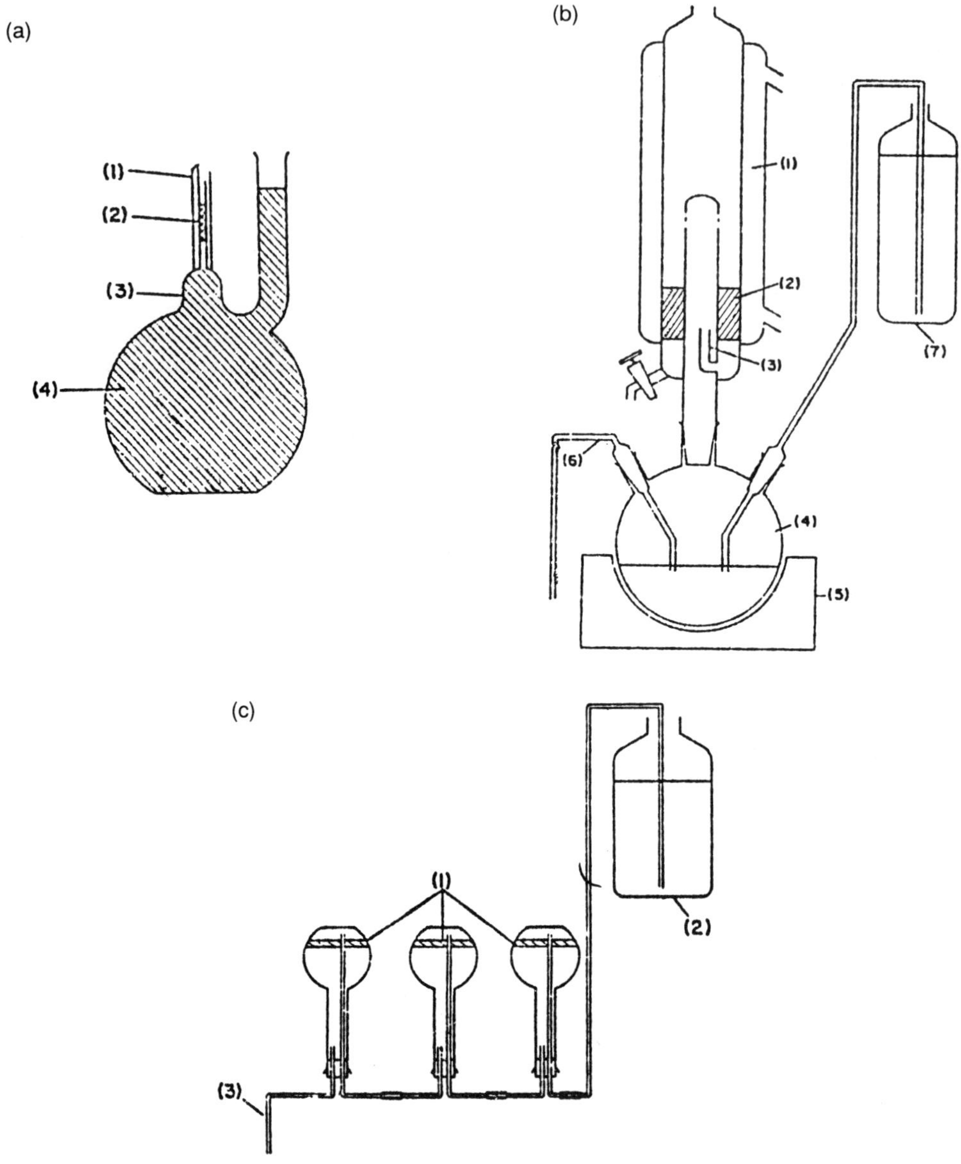

Fig. 4 (a) Micro extraction flask: 1 = capillary tube; 2 = solvent layer; 3 = modified 1-litre volumetric flask; 4 = water sample. (b) Continuous steam distillation and extraction apparatus (modified Nielsen–Kryger): 1 = Condenser; 2 = solvent layer; 3 = water overflow; 4 = boiling flask; 5 = heating mantle; 6 = overflow; 7 = sample reservoir. (c) Continuous solvent extraction apparatus: 1 = solvent layer (hexane), 2 = sample reservoir; 3 = overflow. From Murray[13] with permission.

by Veith and Kiwus[15]) was made with a reduced solvent capacity of 10 ml. Each 10 litre water sample was passed through this system (Fig.4 (b)) in about 8 h. The hexane layer was removed and analysed by gas chromatography.

A continuous extraction apparatus was developed (Fig. 4(c)) which consisted of three inverted 250 ml volumetric flasks clamped to a retort stand, from which a bar was connected to a vibratory mixer. This was adjusted to give a vigorous shaking of all three flasks. Each 10 litre sample was siphoned through the apparatus in about 4 h and extracted with a total solvent volume of 10 ml. The hexane was recovered in a separatory funnel for analysis by gas chromatography. Quantitative analyses are based on the use of internal standards and integration of peak areas. Glass distilled hexane was used throughout.

In order to measure evaporation losses, hexane was spiked with 1 mg litre^{-1} concentrations of three separate groups of organic compounds, concentrated from 10 ml to 1 ml using a micro Snyder column at 90 °C. and a rotary evaporator at 50 °C and analysed by gas chromatography. Recovery of hydrocarbons with boiling points up to 250 °C was low (from 8–16 per cent) with both the micro Snyder column and the rotary evaporator. The recovery of phthalate esters and pesticides using the rotary evaporator was slightly better than the micro Synder column but both techniques showed losses of 10 per cent or more for pesticides (Table 40).

Table 40 Recoveries on concentrating from 10 ml to 1 ml in hexane.

Substance	Boiling point (°C)	Recovery (per cent)	
		Micro Snyder column (90°C)	Rotary evaporator (50°C)
Lindane	–	81	92
Aldrin	–	77	84
Heptachlor epoxide	–	78	90
a-cis-Chlordane	–	78	87
Dieldrin	–	79	85
n-Octane	126	37	8
n-Decane	174	57	25
n-Dodecane	216	59	39
n-Tetradecane	254	61	49
Dipropyl phthalate	305	81	100
Diisobutyl phthalate	298	82	96
Dibutyl phthalate	340	82	100
Butyl glycolyl butyl phthalate	219 at 5 mmHg	84	100

From Murray[13] with permission.

A Hewlett Packard 5750 gas chromatograph was used with an Inotronics C.R.S. 208 inetegrator for quantitative analyis. A 2 m × 6 mm o.d. glass column was packed with 10 per cent Dexsil 400 coated on Chromosorb W AW, 80–100 mesh. The following settings were used: Temperatures: electron capture detector 260 °C, flame ionization detector 230 °C, injector 220 °C, column 235 °C, or 50–200 °C (programmed at 20 °C min^{-1}); pulse interval: 50 μs; flow rates: helium 25 ml min^{-1}, hydrogen 30 ml min^{-1}, air 250 ml min^{-1}, argon 50 ml min^{-1}.

Potable water samples spiked with ng $litre^{-1}$ concentrations of selected pesticides and μg $litre^{-1}$ concentrations of hydrocarbons (C_{10}, C_{12}, C_{14}, and C_{16}) were extracted successively with 200 μl of hexane and analysed by gas chromatography using electron capture and flame ionization detectors, respectively. There was a mean recovery of 58.3 per cent in the first 200 μl extract and 94.3 per cent in three 200 μl extracts. The standard deviations (calculated from 16 values, four analyses of four extracts) for the first extract range from 2.9 to 10.3 per cent with the higher values at the μg $litre^{-1}$ level. (Table 41). Distilled water was extracted in a similar way and analysed for phthalate esters using a flame ionization detector and showed comparable results. The identity of these phthalate esters was confirmed by gas chromatography—mass spectrometry. The typical *m/e* peak at 149 showed the presence of the phthalate residue and the M + 1 peaks at 279 and 337 using the chemical ionization mode, identified the parent compounds.

Table 41 Recoveries from tap water using micro-extraction procedure.

Substance	Concentration	Recovery in extract (per cent)	Rel. stand. dev.[a] (per cent)	Total recovery in extractions (per cent)
Aldrin	10 ng $litre^{-1}$	47.9	5.6	89.4
Heptachlor epoxide		58.5	8.2	91.3
α-*cis*-Chlordane		59.2	10.2	92.0
Dieldrin		62.2	10.0	92.6
n-Decane	50 μg $litre^{-1}$	69.3	3.7	98.6
n-Dodecane		60.3	3.3	97.3
n-Tetradecane		56.0	3.9	96.7
n-Hexadecane		52.9	2.9	96.4
Dibutyl phthalate	2.5 μg $litre^{-1}$	65.5	5.9	96.6
Butyl glycolyl butyl phthalate	25 μg $litre^{-1}$	43.6	4.0	89.7

[a] Four analyses of four extracts.
From Murray[13] with permission.

When 1 litre of potable water was extracted with 200 μl of solvent, only about 50 μl were recovered because of losses due to solubility and evaporation. Thus, a concentration factor of × 20 000 was achieved. But since the extraction efficiency was only about 50 per cent in the first extract the true concentration factor was × 10000. When 10 litres of potable water were extracted with 10 ml of solvent, losses due to solubility and evaporation are proportionally less significant than with the micro-procedure and about 8 ml of solvent were recovered. Thus the concentration factor was about × 1250. The extraction efficiency will be higher than with the micro-procedure but the concentration factor cannot exceed × 1250 without using a solvent concentration step.

The superior concentration factor of the micro method, the absence of a concentration step, and speed of analysis are advantages in routine quantitative analyses of water samples. Preconcentration factors of 20 000 for traces of organics in potable water samples have been achieved by extraction using diethyl ether and petroleum ether (b.p. 30–40 °C).[15,16] One portion of the extract was reduced to a volume suitable for analysis by gas chromatography—mass spectrometry. The other portion was reacted with diazomethane, which converted some components to methyl derivatives which were then amenable to analysis by a computerized gas chromatography—mass spectrometry system. In order to check for possible procedural contamination two solvent blanks (one for each solvent) were processed in the same way as the extracts.

High resolution glass capillary (porous layer open tubular, PLOT) columns were used for the gas chromatographic separation and each concentrated solvent extraction and blank was run on two different columns, one of which had OV-1 (a methyl silicone gum) as a stationary phase, while the other had FFAP (Carbowax 20 M treated with 2-nitroterephthalic acid) as stationary phase.

Sampling:

Collect 20-litre samples in four pre-cleaned 5 litre glass stoppered bottles and store in the dark during transportation to the laboratory. Before taking a sample, rinse the bottle once with the water to be sampled.

Extraction:

Obtain solvent extracts for GC–MS from the samples collected in a specially designed rig (Fig. 5) as follows. After the sample has been transferred to the rig (previously filled with nitrogen) seal the apparatus under a flow of nitrogen. Then acidify the sample to pH 2 by addition of sulphuric acid (50 per cent v/v usually about 30 ml). Stir the sample (6000 rev min^{-1}) for a few minutes, check the pH and add purified petroleum ether (500 ml, b.p. 30–40 °C). Stop the stirrer after about 5 min and allow the solvent layer to separate (about 20 min). Then pressurize

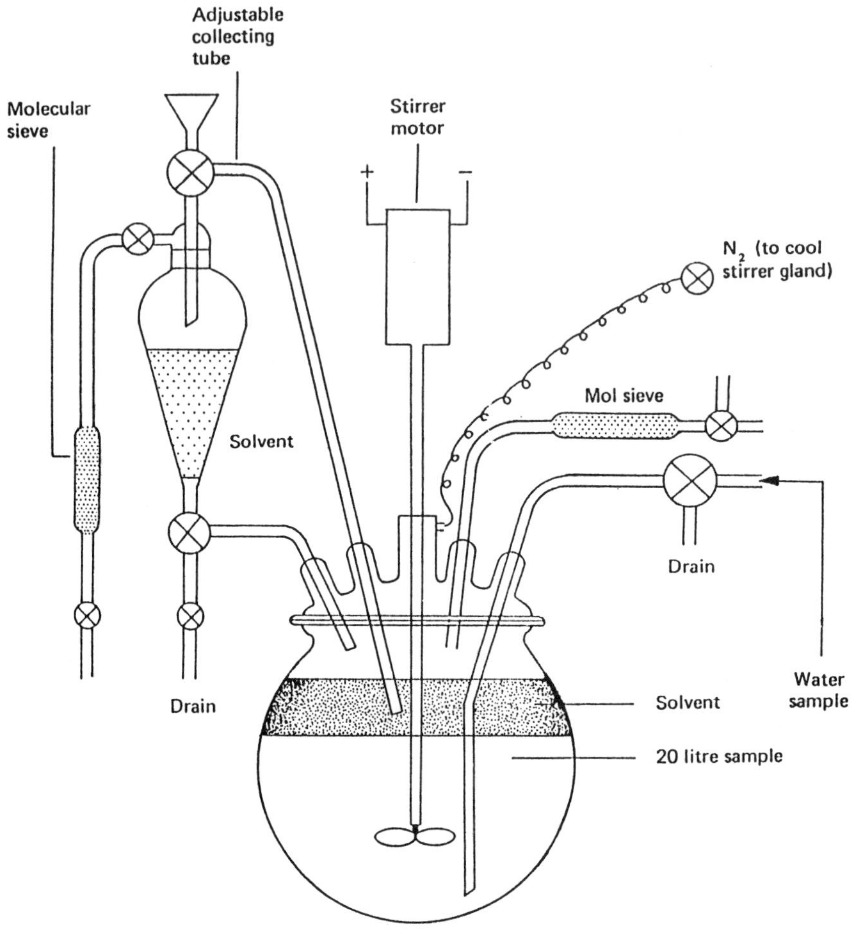

Fig. 5 Solvent extraction apparatus, from Murray[13] with permission.

the solvent layer (using nitrogen via the adjustable take off head and remove it into the separating funnel. Add a second aliquot of petroleum ether (500 ml, b.p. 30–40 ° C) to the sample and repeat the extraction process. Combine the two extracts and dry (pre-treated anhydrous sodium sulphate). For the extraction with diethyl ether, add two aliquots (each of 1 litre) with stirring (5 min) after each addition. Allow the solvent layer to separate and remove from the rig. Due to the solubility of diethyl ether in water, the volume of this first extract (about 500 ml) is considerably less than total amount (2 litres) of diethyl ether added. Add a further aliquot (500 ml) of diethyl ether and extract the sample. Remove the extract from the rig and combine with the first diethyl extract prior to drying (pre treated anhydrous sodium sulphate).

Concentrating:

Concentrate the dried solvent extracts (to 10 ml) on a rotary evaporator, and transfer to precleaned 10 ml centrifuge tubes. Carry out further concentration (to 1 ml) under a stream of dry clean air. Split the extracts equally into two fractions (500 μl) store one fraction (at -18 °C) prior to derivatization, and concentrate the other further (to 100 μl) prior to examination by gas chromatography and mass spectrometry.

Derivatization:

Add a solution (10 ml) of *N*-methyl-*N*-nitrosotoluene-*p*-sulphonamide (100 g litre^{-1}) in diethyl ether:methanol (2:3) to a solution (50 per cent v/v) of potassium hydroxide (1 ml) and distil the diazomethane formed on to the sample to be methylated. Stopper the sample tube and allow to stand for 30 min. Remove the stopper and allow the excess diazomethane to evaporate. Reduce the methylated extract in volume to 100 μl in a stream of dry, clean air.

GC–MS operating conditions:

Hewlett Packard 5710 A (VG 16F).

Carrier gas: helium at 4–5 ml min^{-1}.

Capillary columns: 50 ft or 50 m PLOT columns containing OVI or FFAP (Phase Separations Ltd or Perkin Elmer Ltd).

Injector: SGE splitless injection system SCI-B operated at 250 °C.
Column temperature: Initial 80 °C held for 4 min, then programmed at 8 °C min^{-1} to 250 °C (OV–1) or 220 °C (FFAP) and held at top temperature for at least 10 min.

All spectra were obtained under electron impact (EI) conditions.

Ionisation voltage: 70 eV.

Trap current: 100 μA.

Resolution: 10000.

Source temperature: 220 °C.

Transfer line temperature: 260 °C.

Electron multiplier gain: 2 kV.

Mass range: 20–460 a.m.u.

Scan speed: 1 decade^{-1}.

Interscan delay: 1.

Data system:

The data system used throughout was a VG 2040 system. Perfluorokerosene was used as the calibration compound for all the GC–MS runs. The calibration step was always followed by the acquisition of several mass spectra of perfluorokerosene in order to check the calibration.

Generally, 1000 to 1250 mass spectra were acquired during each GC–MS run. The initial data interpretation step involved an inspection, usually on a visual display unit, of the computer reconstructed total ion current chromatogram and the mass spectra obtained for each obvious peak in the chromatogram was examined. If necessary a background subtraction was performed. If the mass spectrum of a compound of interest was not visually recognized as a known compound, reference was made either to the Eight Peak Index of Mass Spectra[17] or the EPA/NTH Mass Spectral Data Base.[18] If a suitable spectral match was not obtained from either of these sources, the fragmentation pattern of the spectrum of interest was examined and an attempt made to deduce the structure of the compound giving rise to the spectrum. Identifications made either from reference data or from an examination of the mass spectrum produced were considered to be tentative until a pure sample of a known compound gave a suitable match for the mass spectrum and gas chromatographic retention time. Mass chromatography was used to search for compounds of interest which did not appear as obvious peaks on the TLC chromatogram.

Rooney *et al.*[19] extracted miscellaneous organics from natural water using diethyl ether, carbon disulphide or methylene dichloride and achieving preconcentration factors of 10^5. Gas chromatography was used in the analytical finish. Gomez–Taylor *et al.*[20] similarly achieved a preconcentration factor of 2.5×10^5 in diethyl ether extractions of miscellaneous organics from natural water. These workers used Fourier transform infra-red spectroscopy for the analytical finish.

Ibrahim and Suffet[21] used Freon 113 (trichlorotrifluoroethane) as an alternative to methylene dichloride for the preconcentration of organics from chlorinated potable water. Freon minimized the interference from solvent artifacts arising from the reaction of cyclohexene (preservative in methylene chloride) with residual free chlorine (present in chlorine treated water samples). The extraction efficiency of Freon FC-113 was comparable with that of methylene chloride wnen 100–300 ng non-polar organics per litre was extracted and concentrated 1000-fold. No by products were observed when Freon was used to extract chlorinated water.

The preconcentration by solvent extraction of various individual types of organic compounds is now discussed.

5.1.1 Aliphatic hydrocarbons

See Table 42.

5.1.2 Aromatic hydrocarbons

Stoffmeister *et al.*[22] preconcentrated monocyclic aromatic hydrocarbons from potable water by micro-extraction with *n*-pentane and

Table 42 Preconcentration solvents for organics.

Substance preconcentrated	Type of water	Solvent	Finish technique	Ref.
Alkanes	Natural	Pet. ether	GLC	35
		C_5H_{12}	GLC	36,37
		C_6H_{14}	GLC	38,39
		C_6H_{14}	UV	39
		C_8H_{18}	GLC	40,41
		ISO-C_8H_{18}	GLC	42
		$PhCH_3$	GLC	37
		$PhCH_3$–PhC_2H_5	GLC	37
		$(C_2H_5)_2O$	GLC	43,75
		$(C_2H_5)_2O$	Fluorescence	44–47
		CH_3OH–C_6H_6	GLC	48,49
		$C_6H_5NO_2$	GLC	50–52
		CH_2Cl_2	IR	53
		$CHCl_3$	GLC	36,54
		CCl_4	GLC	36,37,50,55–57
		CCl_4	UV	58,59
		CCl_4	IR	50,60–72
		CCl_4	NMR	73
		$C_2Cl_3F_3$	NMR	73
		$C_2Cl_3F_3$	IR	74
Aromatic hydrocarbons	Natural	CS_2	GLC	76
	Rain	$C_6H_5NO_2$	GLC	77
	Rain	C_6H_6	Spectrophoto	78
	Natural	$CH_3COOC_2H_5$	GLC–MS	79

Table 42 cont.

Substance preconcentrated	Type of water	Solvent	Finish technique	Ref.
Polyaromatic hydrocarbons	Natural	n-C_5H_{12}	GLC	80
	Natural	cyclo C_5H_{10}	TLC	81,82
	Potable	cyclo C_5H_{10}	Spectrofluoro	83
	Potable	cyclo C_5H_{10}	GLC	84
	Natural	C_6H_6	TLC	85
	Natural	CH_2Cl_2	TLC	86
	Natural	CH_2Cl_2	GLC	80,86–91
	Natural	CCl_4	TLC	92
	Natural	$(C_2H_5)_2O$	GLC	80
	Natural	$(CH_3)_2CO$	GLC	80
	Natural	Acetonitrile	GLC	80
	Natural	Dimethyl sulphoxide	GLC	86,93
Phenols	Natural	CH_2Cl_2	GLC	94
	Potable	$(C_2H_5)_2O$	Spectrophoto	95
	Natural	$(C_2H_5)_2O$	GLC	96
	Natural	$CH_3COOC_2H_5$	GLC/MS	97
	Natural	$CH_3COOC_2H_5$	GLC	98
	Rain	$C_6H_5CH_3$	GLC	99
Carboxylic acids	Natural	CH_2Cl_2	GLC	100
		$CHCl_3$	TLC	101
		$(C_2H_5)_2O$	Spectrophoto	102
		CH_3OH–$CHCl_3$	GLC	103
		C_6H_6–CH_3OH	GLC	104,105

Chlorinated hydrocarbons	Natural	C_5H_{12}	GLC	106–108
	Lake	C_5H_{12}	GLC	109
	Natural	C_6H_6–C_6H_{14}	GLC–MS	110
	Natural	CH_2Cl_2	GLC	111
	Natural	CCl_4	NHR	112
	Natural	$C_2F_3Cl_3$	NMR	112
Chlorophenols	Natural	n-C_6H_{14}	GLC	113
	Rain	$C_6H_5CH_3$	GLC	114
Polychlorobiphenyls	Natural	C_6H_{14}	GLC	115,129
Haloforms	Potable	Pet. ether	GLC	116
		Pet. ether	HPLC	117
		C_5H_{12}	GLC	118–124
		C_6H_{14}	GLC	118,125
		C_7H_{16}	GLC	118
		ISO-C_8H_{18}	GLC	118–120
		n-C_9H_{20}	GLC	118
		Me cyclohexane	GLC	118,126,127
		$C_6H_4(CH_3)_2$	GLC	128
Chlorinated insecticides	Natural	Pet. ether	GLC	130–134
	Natural	n-C_6H_{14}	GLC	130–133,135–140
	Rain	C_5H_{12}	GLC	141,155
	Natural	$(C_2H_5)_2O$	GLC	140,142,143
	Natural	C_6H_6	GLC	134,144,145
	Natural	CH_2CCl_2:C_6H_{14}(15:85)	GLC	146
	Natural	CH_2Cl_2	GLC	146,147
	Natural	$CHCl_3$	TLC	148

Table 42 cont.

Substance preconcentrated	Type of water	Solvent	Finish technique	Ref.
	Natural	Acetonitrile	GLC	134
	Natural	$C_2F_3Cl_3$	NMR	149
α-Bendo-sulphon	Natural	C_6H_{14}	GLC	150
Vapona	Lake	$CHCl_3$	GLC	151
Trichlorpyr	Natural	$(C_2H_5)_2O$	GLC	152
Dichlorvos	Natural	C_6H_{14} $C_6H_5CH_3$ C_6H_{14}–C_6H_6	GLC	153
Dieldrin	Natural	Pet. ether	GLC	154
Phosphoric acid esters (after hydrolysis)	Natural	$CHCl_3$	GLC	156
	Natural	CH_2Cl_2	GLC/MS	157
Organophosphorus insecticides	Lake	n-C_6H_{14}	TLC	158
	Natural	C_6H_6	GLC Spectrophoto	159,160
	Natural	CH_2Cl_2	GLC	161,162
	Natural	$CHCl_3$	TLC	163
	Natural	$CHCl_3$	GLC	151
Herbicides	Natural	ISO-C_8H_{18}	GLC	164–169

Herbicides carbamate type	Natural	C_6H_6	GLC	164–169
	Lake	CH_2Cl_2	Spectrophoto	170
	Natural	CH_2Cl_2	GLC	162,171
	Natural	CH_2Cl_2	TLC	172–175
m-s-Butyl phenyl methyl (phenylthio) carbamate	Natural	$CHCl_3$	GLC	171
	Natural	Acetonitrile	GLC	171
Phenoxyacetic acid type	Natural	C_6H_{14}	GLC	176
		C_6H_6	GLC	177
		$(C_2H_5)_2O$	Spectrophoto	178
		$(C_2H_5)_2O$	TLC	179
		$(C_2H_5)_2O$	GLC	180
		$CH_3COOC_2H_5$	GLC	180
Triazine type	Natural	CH_2Cl_2	GLC	181–183
Urea type	Natural	$C_6H_5CH_3$	GLC	184
		CH_2Cl_2	GLC	185–186
		CH_2Cl_2	TLC	172
		$CHCl_3$	GLC	184
		CCl_4	GLC	180
		$(C_2H_5)_2O$	GLC	184
		CH_3NO_2	GLC	184
Misc. types	Natural	CH_2Cl_2	GLC	187
Dalapon	Natural	$(C_2H_5)_2O$	TLC	188
Paraquat, diquat	Natural	CH_2Cl_2	GLC	189
Dicamba	Natural	$(C_2H_5)_2O$	GLC	190

Table 42 cont.

Substance preconcentrated	Type of water	Solvent	Finish technique	Ref.
Nitrosamines	Natural	CH_2Cl_2	GLC	191,192
Non-ionic detergents as tetrathiocyanato cobalt(II) complex	Lakes	$C_6H_5CH_3$	AAS	193
Chloroanisoles	Natural	CH_2Cl_2	GLC	194
Pyrethrins	Potable	C_6H_{14}	GLC	195
Amines	Potable	C_6H_{14}	GLC–MS	196
Siloxanes	Natural	Pet. ether	ICPAES	197
Vegetable oil	Natural	CCl_4	IR	198
Linear alkyl benzene sulphonates	Natural	C_6H_{14}	GLC	199
Geosmin	Natural	CH_2Cl_2	GLC/MS	200
		CH_2Cl_2–*n*-tridecane	GLC	201
Squoxin	Natural	*n*-C_3H_7OH–CCl_4 (2:1)	spectrophoto	202
ETDA	Natural	$CHCl_3$	GLC	203,204
Nitrophenols	Natural	$C_6H_5CH_3$	GLC	205

addition of di-isobutylene as an internal standard before determination by gas chromatography with a flame ionization detector. The detection limit was in the range 0.1–0.5 μg $litre^{-1}$. See also Table 42.

5.1.3 Polyaromatic hydrocarbons

Concawe[23] recommends methods described by the Environmental Protection Agency for the determination of polyaromatic hydrocarbons in oil refinery effluents. The method involves extractive preconcentration of the effluent with methylene dichloride followed by clean-up procedures followed by gas chromatography or liquid chromatography.

For the gas chromatographic method no detector is specified; flame ionization seems to be the best choice. Lack of selectivity in the method can lead to interference by compounds that are not completely removed by the clean-up. Higher selectivity can be achieved by use of a photoionization detector. In addition some pairs of the polyaromatic hydrocarbon isomers are incompletely separated by the 6′ column used in this method while the heavier polyaromatic hydrocarbons often show tailing peaks. Almost all polyaromatic hydrocarbon isomers can be separated using a longer column (30′) with the same packing material, or by using a glass capillary column; only benzo(b)fluoranthrene–benzo(k)fluoranthene are not resolvable even by the long packed, or capillary column approach, both of which are recommended as a final step in preference to a conventional colum. The liquid chromatographic procedure[24] uses reversed phase liquid chromatography with fluorescence detection to separate all 16 polyaromatic hydrocarbons completely. The method is sensitive and so selective as often to allow the method to be applied without clean-up procedure.

For gas chromatographic methods detection limits are about 1 μg $litre^{-1}$ whereas for the liquid chromatographic methods limits are between 1 and 100 μg^{-1} for 2-and 3-ring aromatics and below 1 μg $litre^{-1}$ for the 4-, 5-, and 6-ring compounds. For polyaromatic hydrocarbon analysis the extraction with methylene chloride may successfully be substituted by cyclohexane allowing high recoveries by one single extraction. Isolation of polyaromatic hydrocarbons from the water matrix may also conveniently be performed via adsorption using prepacked small high performance liquid chromatography columns circumventing the problems of emulsification often arising during liquid–liquid extraction. Recently, this technique has been used in a study of polynuclear aromatic hydrocarbons in aqueous effluents from refineries as it was shown to differentiate between certain compounds which were not resolved by gas chromatography—mass spectrometry.

See also Table 42.

5.1.4 Phenols

Czuczwa *et al.*[25] preconcentrated traces of phenols and cresols from rain water by a method based on continuous liquid–liquid extraction (solvents heavier than water) and normal phase and high performance liquid chromatography. See also Table 42.

5.1.5 Carboxylic acids

Fataki and Vernon[26] used chloroform extraction followed by capillary column gas chromatography to estimate extremely low levels of fatty acids in river waters and sewage effluents.

5.1.6 Chlorinated hydrocarbons

See Table 42.

5.1.7 Chlorophenols

Fingler *et al.*[27] carried out a comparative study of trace enrichment of chlorophenols from ground and potable waters by extraction with C_6 and C_8 alkanes and by octadecyl carbon C18 reversed phase adsorption. All three procedures could efficiently enrich traces of chlorophenols, the choice of procedure depending on the required analytical sensitivity. Extraction procedures could be applied to concentrations of chlorophenols at or above the $1\,\mu g$ litre^{-1} level. However, the C18 reversed-phase adsorption prodecure achieved a detection limit of 10 ng liter^{-1}. See also Table 42.

5.1.8 Polychlorobiphenyls

Kahn and Wayman[28] reported a continuous multichamber liquid–liquid extractor with internal solvent refreshing for the extraction of non-polar contaminants, especially polychlorobiphenyls and chlorinated pesticides from natural waters. With this apparatus, it was possible to extract 135 litres of water at rates of 0.5 to 1.0 litre h^{-1} and in that way isolate and identify μg litre^{-1} concentrations of these compounds with a recovery of 97 per cent. Goldberg *et al.*[29] presented a similar apparatus which was capable of handling flow rates up to 2 litres h^{-1} designed to operate with heavier than water solvents. In a later work Goldberg *et al.*[30] described modified extractors for both heavier than water and lighter than water solvents. Three solvent heavier than water extractors were set up in one series and four solvent lighter than water extractors were set up in a second series. Spiked water was pumped at a rate of 7 to 8 litres h^{-1}

through each set of extractors and the extraction efficiencies were determined. A concentration factor of up to 10^5 was obtained and the dipole moment difference between the solute and solvent was demonstrated to be an index of the extraction efficiency.

Ahnoff and Josefsson[31] modified the solvent extraction apparatus of Kahn and Wayman,[28] to fit field conditions by using one chamber and enlarging the outlet tube. The phase separation became more efficient and the stirring rate could be increased without losing drops of the organic solvent. This arrangement is similar to that described by Kerner *et al.*,[32] who used a bulb for better phase separation. However, a system where the organic solvent is recycled is too complicated for routine field application, especially when immersed in the water. The changes of the outer temperature will disturb the heating of the solvent. The few theoretical plates in the column between the evaporation flask and the reflux condenser did not prevent the organic compounds cycling together with the solvent. Water is dissolved in the vaporizing flask. The boiling of the solvent together with water causes, for example, uncontrollable heat changes, hydrolysis reactions, and steam distillation. In the system devised by Ahnoff and Josefsson[31] the organic refreshing system normally used was eliminated. Chlorinated biphenyls were extracted from some hundred litres of water at the Gota river with different multi-chamber arrangements. Finally, the content of polychlorinated biphenyl was determined with ECD gas chromatography in the range 0.1–1.0 ng $litre^{-1}$ of water.

Apparatus:

The apparatus is shown in Fig. 6. In principle, it is a mixing settler although the mixing and settling chambers are not completely separated but combined in one cylinder. The water is continuously drawn through the cylinder under vigorous mixing with a lighter than water organic solvent, thereby forming a vortex. The mixing is performed by a magnetic stirrer (A) located under the cylinder. The organic solvent is trapped by a flange in the cylinder and is not replenished during the procedure. The mixing of the water and solvent takes place in the upper part of the cylinder and the emulsion is separated successively in the lower part. The lighter solvent droplets are allowed to rise and return into the mixing zone. In this way, very little of the extracting solvents is lost during the procedure. Use of a solvent volume of 100–300 ml with the 800 ml apparatus (shown in Fig. 6 (a)) is recommended.

The apparatus consists of three main parts; the extraction unit, the magnetic stirring device (A), and the pump (B) (ProMinent Electronic 0304T, CFG, West Germany). The extraction unit is designed in four parts; top (C), bottom (D), bowl (E) all in PTFE and a borosilicate glass cylinder (F), 80 mm i.d. by 190 mm, and a wall thickness of 5 mm. The

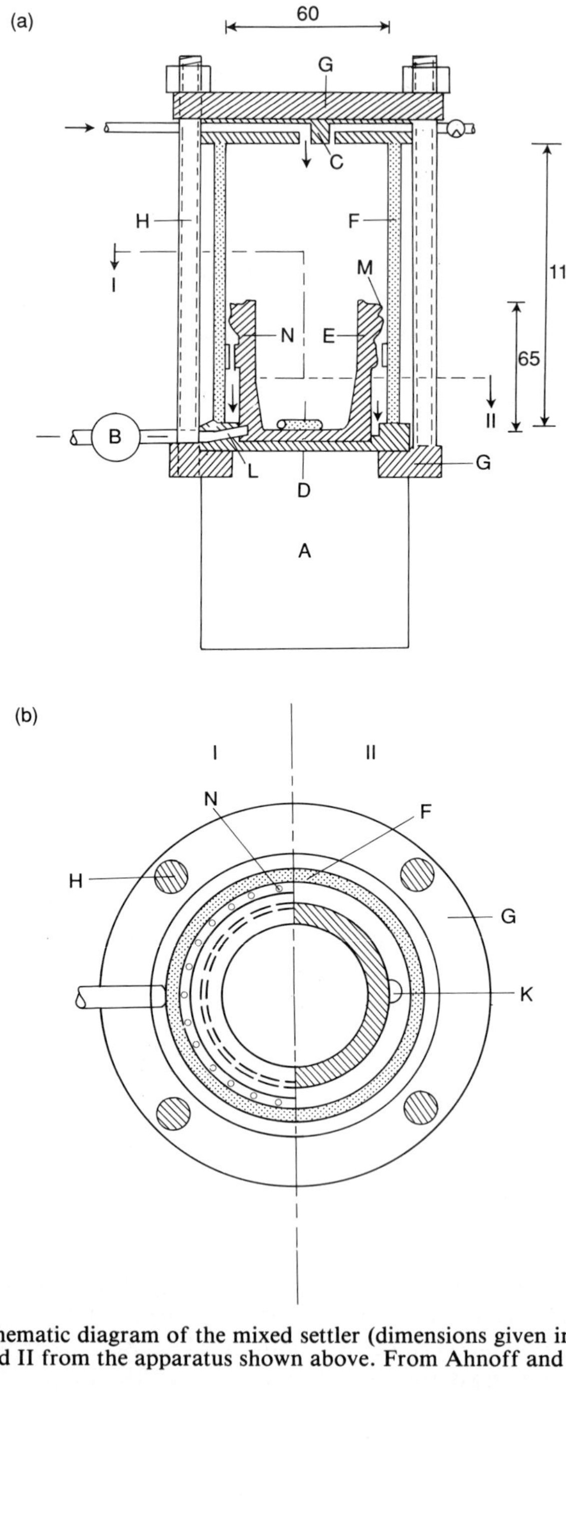

Fig. 6 (a) Schematic diagram of the mixed settler (dimensions given in mm). (b) Cross section at I and II from the apparatus shown above. From Ahnoff and Josefsson[31] with permission.

top and bottom parts are pressed water tight to the glass cylinder by plates of acid proof stainless steel (G) and four screwbolts (H). The bottom steel plate has an opening in the middle (65 mm in diameter) to actuate the stirring bar inside (I). The magnetic stirring bar (28 mm × 18.5 mm) rotates at about 900 r.p.m. In the top part are two channels (3 mm in diameter) one for the water inlet and one, equipped with a stopcock, for the filling and the draining of the apparatus. The bottom part (D) has a circular hollow to fit the bottom of the bowl (E). Three symmetrically placed holes (K) at the edge of the hollow are in contact with a circular channel (L) which leads to the outlet tube (B). In this way, the extracted water is sucked out from the calm circular compartment (O) with a symmetrical flow. The bowl (E) is designed to keep the mixing process above and inside whereas the outside of the bowl acts to separate the emulsion and return the organic solvent into the mixing zone. The bowl has a double flange (M) to quell the vertical emulsion movement. The circular cross section area between the flange and the glass wall is important because the flow rate acts to bring the emulsion downward while the lighter solvent is striving upward. The rotating movement in the upper compartment is definitely quelled at N (Figs 6(a) and (b)) by letting the water pass 36 holes (3.6 mm in diameter). The final phase separation occurs in the cavity (0) below the holes where the mixing procedure has no influence. By this arrangement of the settling zones, most of the ligher solvent is trapped and returns to the mixing compartment, provided the flow rate is not too high (see below). Since there is no pressure drop in the apparatus, the pump can be placed at the outlet. Thus contamination caused by the pump is eliminated.

Materials:

The cyclohexane used as extracting agent was purified by distilling technical grade cyclohexane on a bubble cap plate column with 12 plates, using a reflux ratio of 12:1. The purity was checked by concentrating 200 ml of the solvent to 1 ml and injecting 10 μl into the gas chromatograph. *n*-Hexane 85 per cent (Merck) was distilled and checked in the same way as described for cyclohexane. The standard pesticide solution used in the laboratory extraction tests contained 5 μg lindane, 10 μg aldrin, 25 μg dieldrin, 20 μg *p,p'*-DDE, 40 μg *p,p'*-DDD, and 60 μg *p,p'*-DDT (all 99.9 per cent pure) per ml of acetone, AR (Merck). Clophen A-60 (Bayer) a mixture of polychlorinated biphenyls (PCB) with 60 per cent chlorine content was used as a gas chromatographic standard for the polychlorinated biphenyl determination in the river water extracts. In the clean-up procedure employed for the polychlorinated bipheyl analysis, Florisil 60–100 mesh (Metheson Coleman & Bell) and sodium sulphate AR (Merck) were used. Both reagents were heated to 600 °C in 4 h and kept at 130 °C in glass stoppered vials.

Procedure:

Set up three extractors in one series connected by PTFE tubing. Charge each extractor with 150 ml of cyclohexane and fill up with triple distilled water. In a 10 litre glass flask, which serves as a feeding reservoir, introduce 8 litres of triple distilled water and 1 ml of standard pesticide solution 10 min before the extractions start. Mix the spiked water throughout the whole experiment with a magnetic stirrer. Connect the outlet of the third extractor to a pump which sucked the water from the feeding reservoir though the three extractors. Collect the effluent water in another 10 litre glass flask. After the 8 litres had been sucked through the system, add 1 litre of triple distilled water twice to the feeding reservoir and pump through the system, the extractors still being in action. This was done to ensure that all of the spiked water had passed through all three extractors. After the extraction, turn the three extractors upside down and drain. Examine the extracts on the gas chromatograph. Shake the feeding reservoir, the drained first extractor, and the effluent water vigorously with 100 ml of cyclohexane. Also analyse these samples. Use the standard pesticide solution to spike the water and also to prepare the gas chromatographic standards. Thus the amounts of pesticides found in the different samples were determined as fractions of the amount added to the feeding reservoir.

Field procedure:

Place the extractors on the river bank near the water surface. The pump sucks the river water unfiltered through the extractors. Then arrange the extractors to extract in series and parallel extractor trains. Before use, thoroughly clean the extractors. To check the background of interfering contaminants deriving from the extractor as well as the clean-up procedure prepare a blank from each extractor. Fill the extractors with 250 ml of cyclohexane and triple distilled water and stir the contents vigorously for 24 h. Then treat the cyclohexane in the same way as the river water extracts.

Treatment of extract before gas chromatography:

Inject the extracts froin the test procedure into the gas chromatograph without previous treatment. Concentrate the extracts from the field procedure to 5 ml in a Kuderna Danish concentrator and a tube heater (Kontes, K 720000). Place the concentrate on a 10 cm Florisil column (2 cm i.d.) covered with 2 cm of sodium sulphate, and elute with 200 ml of hexane (the column had been prewashed with 100 ml of hexane). Concentrate the eluate to 1 ml in the same apparatus as above and inject 10 μl portions into the gas chromatograph. Control the recovery of polychlorinated biphenyl by running 200 ng of Clophen A 60 in 5 ml hexane through the clean up and concentration procedures.

Gas chromatography:

A Hewlett Packard Model 7620 A dual column gas chromatograph equipped with an electron capture detector (200 mCi tritium; pulsed system) was employed. Glass columns (2 m × 2 mm i.d.) silanized with dimethyldichlorosilane, were used. The solid suppport was 80/100 mesh Chromosorb W DMCS coated with 3 per cent OV-1, 3 per cent QF-1, and 3 per cent OV-17/3 per cent XE–60 (1:1) respectively.

The conditions for the chromatography were: injector, 230 °C; column 195 °C; detector, 215 °C. Helium was used as the carrier gas at a flow rate of 40 ml min^{-1}. The purge gas was argon with 10 per cent methane using a flow rate of 60 ml min^{-1}. The electron capture detector pulse interval was 50 μs and the standing current on the chromatograph was held steady during the analysis at 5×10^{-9} A. The peak height was used for the quantitative evaluation. Dieldrin and *p,p'*-DDE were not separated on the OV-1 column; accordingly, their sum was reported. The quantitative calculation of the PCB was made on the five highest peaks in the standard. The peak height sums in the sample and the standard were calculated and compared.

Table 43 presents the amounts of polychlorinated biphenyls found in the river water extracts from different field experiments at the Gota river. Three different extraction arrangements were set up in five

Table 43 Concentration of PCB found in Göta River water at *c.* 1 m.

Arrangement	Date	Water volume (litres)	Pump rate (litre h^{-1})	PCB (ng $litre^{-1}$)
Parallel	8 Sept. 1972	290	5.0	0.31
	8 Sept. 1972	305	5.2	0.29
Parallel	17 Jan. 1973	170	3.5	0.21
	17 Jan. 1973	185	3.8	0.22
Series 1				
Extractor 1	29 Jan. 1973	190	3.8	0.32
Extractor 2	29 Jan. 1973	190	3.8	0.16
Series 2				
Extractor 1	31 Jan. 1973	170	3.4	0.20
Extractor 2	31 Jan. 1973	170	3.4	0.12
Series 3				
Extractor 1	3 June 1973	240	3.0	0.28
Extractor 2	3 June 1973	240	3.0	0.17
Extractor 3	3 June 1973	240	3.0	0.09

From Ahnoff and Josefsson[31] with permission.

experiments; two parallel extractor experiments and three experiments with the extractors arranged serially. The water flow varied as well as the total amount of water pumped through the system.

Petrick *et al.*[33] used hexane to preconcentrate polychlorobiphenyls from natural water prior to analvsis by gas chromatography. Aliphatic and polyaromatic compounds were first separated from the extract by high performance liquid chromatography. See also Table 42.

5.1.9 Haloforms

Vartiainen *et al.*[34] used diethyl ether liquid–liquid extraction to preconcentrate mutagenic organic compounds from chlorinated potable water with a high humus content. See also Table 42.

5.1.10 Chlorinated insecticides

See Section 5.1.8. and Table 42.

5.1.11 Phosphoric acid esters

See Table 42.

5.1.12 Organophosphorus insecticides

Leoy[35] preconcentrated azinphos methyl, fenthion, and diazinon by extraction from aqueous solution with *n*-heptane, isoamyl alcohol, cyclohexane, diethyl ether, dichloromethane or octanol. Diazinon was not efficiently extracted. Calibration graphs were linear for 0.5–0.7 and 8–20 mg $litre^{-1}$ for azinphos methyl (extractions were 90 and 70 per cent, respectively), and up to 4 mg per litre for fenthion (extraction efficiency 33 per cent). Detection limits and relative standard deviation were 0.04 and 0.09 mg $litre^{-1}$ and 3.4 and 5.3 per cent for azinphos methyl and fenthion respectively. Other pesticides and related compounds did not interfere. See also Table 42.

5.1.13 Herbicides

See Table 42.

5.1.14 Miscellaneous organics

See Table 42 for a review of the following types of organic compounds; nitrosamines, non-ionic detergents, chloroanisoles, pyrethrins, amines, siloxanes, vegetable oils, linear alkyl benzene sulphonates, geosmin, squoxin, EDTA, and carbon disulphide.

5.2. Organic compounds—sea-water

5.2.1 Sampling techniques

Solvent extraction is an attractive method for preconcentrating a particular fraction of the dissolved organic matter in sea-water, the fraction concentrated being determined by the choice of solvent. The most obvious limitation on the method is that set by solvent choice; the solvent should have only limited solubility in water, which limits the materials removed to the less polar compounds. Another limitation is that set by contamination. The solvent used must be purified carefully, since the amounts of the various organic compounds collected from sea-water will be about as large as the trace impurities in the solvents. Because of the relatively large amount of time and equipment required for the processing of each sample, these methods must be used to characterize the organic compounds at a few selected stations and depths, rather than in mass surveys. Once the separation into the organic solvent has been accomplished, any of a number of techniques of fractionation and analysis can be applied.

The first step in the solvent extraction is the actual sampling of the water column. All of the problems associated with sampling can occur in this step, and may be aggravated by the large volumes of sea-water customarily employed. The cleverest approach to this problem is to avoid sampling in the normal manner altogether. Ahnoff and Josefsson[206] have described an *in-situ* apparatus for solvent extraction. This apparatus is buoyed at the sampling depth, anywhere between the surface and 50 m, and water is pumped through a series of extraction chambers. The capacity of the unit is 50 litres per 48 h. The use of an *in-situ* pumping system on the far side of the extraction chambers eliminates the pump and hose contamination, as well as much of the contamination coming from the passage through the surface film. Since the apparatus is battery powered, the unit may be suspended from a free-floating surface buoy; no ship time is required, except for placement and recovery of the samplers. Thus while each sample may require up to 48 h to collect, a number of depths and areas may be sampled in the same time period.

Since the non-polar organic content of sea-water is fairly low, identification of unknown compounds requires the processing of large quantities of sea-water. Where possible, continuous extraction should be favoured over batch extraction. Several workers have developed such systems: Goldberg *et al.*[207] designed and evaluated a system for fresh water. This apparatus seems fragile, and would require considerable redesign before it could be used routinely on shipboard. Ahnoff and Josefsson[208] built a solvent extraction apparatus for river work which was later modified into their *in-situ* extractor.[209] The unit as described in the earlier work could easily be adapted for sea-water analysis. A unit based on a Teflon helix liquid–liquid extractor, some 332 feet (101.5 m) in length, was constructed by Wu and Suffet.[210] The extractor was optimized for the removal of organophosphorus compounds, specifically pesticides, with an efficiency of around 80 per cent. For some compounds, these continuous extraction methods should be the methods of choice, and should be further explored. These compounds are the ones that are not seriously affected by hose or pump contamination, are present in low concentrations, and are important enough, for either scientific or practical reasons to warrant the extra time spent on station.

The conventional approach to solvent extraction is the batch method. Early work with this method was hampered by the low concentrations of the compounds present and the relative insensitivity of the methods of characterization. Thus lipids and hydrocarbons have been separated from sea-water by extraction with petroleum ether and with ethyl acetate. The fractionation techniques included column and thin-layer chromatography with final characterization by thin-layer chromatography, infra-red and ultraviolet spectroscopy, and gas chromatography. Of these techniques, only gas chromatography is really useful at the levels of organic matter present in sea-water. With techniques available today such as glass capillary gas chromatography and mass spectrometry, much more information could be extracted from such samples.[211]

The information could be restricted to a tractable amount by performing some preliminary fractionation before the gas chromatography—mass spectrometry step. This type of separation and fractionation has been proposed by Copin and Barbier.[212] During the investigation of pollution in coastal sea-waters, Werner and Waldichuk[213] pointed out the need for concentrating and isolating trace amounts of certain substances with a continuous solvent extractor. They constructed a modified Scheibel[15] apparatus by changing the organic solvent cycle system.

5.2.2 *Hydrocarbons*

Zsolnay and Kiel[214] have used flow calorimetry to determine total hydrocarbons in sea-water. In this method the sea-water (1 litre) was extracted with trichlorotrifluorethane (10 ml) and the extract was concentrated, first in a vacuum desiccator, then with a stream of nitrogen to 100μl. A 50 μl portion of this solution was injected into a stainless steel column (5 cm × 1.8 mm) packed with silica gel (0.063–0.2 mm) deactivated with 10 per cent of water. Elution was effected, under pressure of helium, with trichlorotrifluorethane at 5.2 ml^{-1} and the eluate passed through the calorimeter. In this the solution flowed over a reference thermistor and thence over a detector thermistor. The latter was embedded in porous glass beads on which the solutes were adsorbed with evolutions of heat. The difference in temperature between the two thermistors was recorded. Each solute first displayed the adsorbed solvent molecules, giving a desorption peak (negative); this was followed by an adsorption peak (positive). The area of the desorption peak was proportional to the amount of solute present. All hydrocarbons were eluted as one peak, the silica gel column removing the non-hydrocarbon solutes. Responses to unsaturated hydrocarbons were less than those to saturated ones; if the peak area was calculated as nonocane, an error of 25 per cent could be introduced. The limit of detection was about 5 μg litre^{-1}.

Law *et al.*[215] used pentane, hexane, and methylene chloride to extract hydrocarbons from coastal sea-waters. The use of extraction solvents of similar polarity (i.e. pentane or hexane) give comparable results (average variation 22–23 per cent) for samples with total hydrocarbon concentrations in the range 1–2 μg Ekofisk crude oil equivalents per litre. Results obtained using dichloromethane for extraction were higher by a factor of around 2. Probable reasons were: the greater extraction efficiency of dichloromethane owing to its higher polarity; more efficient scavenging of adsorbed hydrocarbons by dichloromethane owing to its higher water miscibility.

5.2.3 Halogenated hydrocarbons

Eklund *et al.*[216,217] have developed a capillary column method[218,219] for the determination of down to 1 μg litre^{-1} volatile organohalides in waters which combine the resolving power of the glass capillary column with the sensitivity of the electron capture detector. The eluate from the column is mixed with purge gas of the detector to minimize band broadening due to dead volumes. This and low column bleeding give enhanced sensitivity. Ten different organohalides were quantified in sea-water. Using this technique these workers detected bromoform in sea-water for the first time. Halogenated hydrocarbons in different waters were identified by comparison with a standard solution. Retention times were measured on two columns with different stationary phases, i.e. SE-52 and Carbowax 400. Extraction of 100 ml water with various amounts of pentane ranging from 1 to 15 ml showed that the extraction efficiency was increased when using a lower water to pentane ratio.

5.2.4 Chlorinated insecticides

Since petroleum ether was the solvent used in earlier studies for extracting the DDT from sea-water, Wilson and Forester[220] initiated further studies to evaluate the extraction efficiencies of other solvent systems. The recovery rates of *o,p'*-DDE in all tests were greater than 89 per cent with petroleum ether, 15 per cent diethyl ether in hexane followed by hexane or methylene dichloride, indicating no significant loss during analyses. The average percentage recovery of *p,o'*-DDT extracted from sea-water containing 3.0 μg litre^{-1} of the DDT isomer (salinity 16–21 ppt) or distilled water samples up to 14 days after initiation of the experiment was of a similar order.

5.2.5 Miscellaneous organics

The Oil Companies International Study Group for Conservation of Clear Air and Water-Europe (Concawe)[221] have made a detailed study of the application of solvent extraction to the determination of organics in water. In this procedure, one portion of the aqueous sample is adjusted to pH 11 and extracted with pure methylene dichloride. The methylene dichloride extract containing basic and neutral substances is examined by gas chromatography coupled to a mass spectrometer. The alkaline aqueous phase is acidified and extracted with methylene dichloride to provide an extract containing phenols which, again is examined by gas chromatography—mass spectrometry. A second portion of the water sample is acidified and extracted with methylene dichloride to provide a further phenol-containing extract for examination by high performance liquid chromatography. A third portion of the water sample is extracted with methylene dichloride and examined for polynuclear aromatic hydrocarbons by high performance liquid chromatography using an ultraviolet/fluorescence detector and by gas chromatography using a flame ionization detector.

Solvent extraction has proved to be most useful when applied to the concentration of particular compounds for which there exists an analytical method of great sensitivity. The major application of the method has been for the determination of hydrocarbons in sea-water. In general, solvent extraction is an excellent method for the concentration and determination of specific compounds, chiefly non-polar, in sea-water.[222] Special precautions must be taken to prevent contamination from trace materials in the solvents used. *In-situ* methods offer many advantages, not the least being the elimination of lengthy processing in the shipboard laboratory; these methods should be investigated more thoroughly and, if possible, extended to greater depths. When coupled to modern separation and detection systems, the methods may offer us the simplest and most direct approach to the measurement of certain classes of compounds. In most cases, we have little or no estimate of the efficiency of the solvent extraction techniques. Because of the great variety of compounds present in any one sample, a true efficiency of extraction may be impossible to obtain. Working efficiencies, using model compounds, may be the only approach in trying to make the analysis truly quantitative.

Hon-Nami and Hanya[223,224] used chloroform extraction to extract linear alkyl benzene sulphates from sea-water prior to analysis by gas chromatography–mass spectrometry.

5.3 Organometallic compounds

5.3.1 Mercury

Non-saline waters Methyl-, ethyl-, and phenyl-mercury compounds have been preconcentrated from natural waters by extraction with benzene, followed by conversion to their cysteine complexes and gas chromatography.[225,226] Pentane:diethyl ether (1:1) has also been used to preconcentrate dialkylmercury compounds prior to analysis by gas chromatography.[227]

Sea-water Organomercury compounds have been preconcentrated from sea-water by forming the dithizonates and extracting these with a small volume of carbon tetrachloride. Mercury is then back-extracted from the organic phase with a small volume of hydrochloric acid and determined by stannous chloride reduction and cold vapour atomic absorpton spectrometry.[228]

5.3.2 Tin

Non-saline waters Various solvents have been used to preconcentrate organotin compounds from natural waters including benzene–tropolone,[229,231] *n*-hexane–tropolone,[230] chloroform–tropolone,[231] methylene dichloride,[231] and benzene.[233] The concentrated organotin compounds are analysed either by direct gas chromatography[229–232] or are converted to tin hydrides and gas chromatographed.[233]

Sea-water Laughlin *et al.*[234] analysed chloroform extracts of tributyltin dissolved in sea-water using nuclear magnetic resonance spectroscopy. It was shown that an equilibrium mixture occurs which contains tributyltin chloride, tributyltin hydroxide, the aquo complex, and a tributyltin carbonate species. Fluorimetry has been used to determine triphenyltin compounds in sea-water.[235] Triphenyltin compounds in water at concentrations of 0.004 to 2 mg litre^{-1} are readily extracted into toluene and can be determined by spectrofluorimetric measurements of the triphenyltin–3-hydroxyflavone complex.

5.3.3 Lead

Non-saline waters Chau *et al.*[236] extracted organolead compounds from water into benzene, prior to conversion to their dithiocarbamates and subsequently to butyl-lead derivatives prior to gas chromatography. Hexane has also been used[237] to preconcentrate alkyl-lead compounds from natural water prior to analysis by gas chromatography using an atomic absorption detector. Down to 0.5 μg litre^{-1} organolead compounds can be determined by this procedure.

References

1. Goldberg, M. C., LeLong, L., and Sinclair, M. (1973). *Analytical Chemistry*, *45*, 89.
2. Peters, T. L. (1982). *Analytical Chemistry*, *54*, 1913.
3. Baker, R. J., Gibs, J., Meny, A. S., and Suffet, L. H. (1987). *Water Research*, *21*, 179.
4. Gomella, C., Belle, J. P., and Auvray, J. (1976). *Techniques et Sciences Minicipals*, *71*, 439.
5. Fritz, J. S. (1977). *Accounts of Chemical Research*, *10*, 67.
6. Kawahara, F. K., Eichelberger, J. W., Reid, B. H., and Stierly H. (1967). *Journal of the Water Pollution Control Federation*, *39* 572.
7. Ahnoff, M. and Josefsson, B. (1976). *Analytical Chemistry*, *48*, 1268.
8. Stepan, S. F., Smith, J. F., Flego, U., and Renkers, J. (1978). *Water Research*, *12*, 447.
9. Goldberg, M. C., LeLong, L., and Kahn, L. (1971). *Environmental Science and Technology*, *5*, 161.
10. Burchill, P., Herod, A. A., March, K. M., Pirt, C. A., and Pritchard, E. (1983). *Water Research*, *17*, 1891.
11. Theron, S. J. and Hassett, D. W. (1986). *Water South Africa*, *12*, 31.
12. Eichelberger, J. W., Kerns, K. H., Olynk, P., and Budde, J. L. (1983). *Analytical Chemistry*, *55*, 1471.
13. Murray, D. A. J. (1979). *Journal of Oceanography*, *117*, 135.
14. Grob, K., Grob, K., and Grob, G. (1975). *Journal of Chromatography*, *106*, 299.
15. Veith, G. D. and Kiwus, L. M. (1977). *Bulletin of Environmental Contamination and Toxicology*, *17*, 631.
16. Fielding, M., Gibson, T. M., James, H. A., McLoughlin, K., and Steel, C. P. (1981). In, *Water Research Centre Technical Report TR 159 Organic micropollutants in drinking water*. February.
17. Mass spectrometry data centre (1974). *Eight peak index of mass*

spectra, 2nd edn, MSDC, Automic Weapons Research Establishment, Aldermaston, Reading.
18. Heller, S. R. and Milne, G. W. A. (1978). In, *EPA/NIH Mass Spectral Data Base*, US Department of Commerce/National Bureau of Standards.
19. Rooney, T. A., Freeman, R. R., Hewlett Packard Ltd, Avondale, US. (1977). In, *Technical Paper No 69. Analysis of organic contaminants in surface water using high resolution gas chromatography and selective detectors*, 174th Meeting of the American Chemical Association National Meeting, Chicago, Illinois, 28 August–2 September.
20. Gomez-Taylor, M. M., Kyehl, D., and Griffiths, P. R. (1978). *International Journal of Environmental Analytical Chemistry*, *5*, 102.
21. Ibrahim, E. A. and Suffett, I. H. (1988). *Journal of Chromatography*, *454*, 217.
22. Stoffmeister, E., Hendel, P., and Hermanan, H. (1988). *Chemosphere*, *17*, 801.
23. The Oil Companies International study group for conservation of clean air and water (1982). In, *Europe CONCAWE report No 6/82. Analysis of trace substances in aqueous effluents from petroleum refineries.*
24. Environmental Protection Agency EPA. 610 Environmental Protection Agency USA.
25. Czuczwa, J., Levenberg, C., Tremp, J., Giger, W., and Ahlel, M. (1987). *Journal of Chromatography*, *403*, 233.
26. Fataki, O. S. and Vernon, F. (1989). *Water Research*, *23* 123.
27. Fingler, S., Porevenkar, U., and Vasilic, Z. (1987). *Mikrochimica Acta*, *No 416*, 163.
28. Kahn, L. and Wayman, C. H. (1964). *Analytical Chemistry*, *36*, 1340.
29. Goldberg, M. C., Delong, L., and Kahn, L. (1971). *Environmental Science and Technology*, *5*, 161.
30. Goldberg, M. C., Delong, L., and Sinclair, M. (1973). *Analytical Chemistry*, *45*, 89.
31. Ahnoff, M. and Josefsson, B. (1974). *Analytical Chemistry*, *46*, 658.
32. Kerner, I., Goto, M., and Korte, F. (1972). *International Journal of Environmental Analytical Chemistry*, *2*, 57.
33. Petrick, G., Schulz, D. E., and Duiker, J. C. (1988). *Journal of Chromatography*, *435*, 241.
34. Vartiainen, T., Lizmatainen, A., Jaaskelainen, S., and Kauranen, P. (1987). *Water Research*, *21*, 773.
35. Leoy, E. M. (1971). *Water Research*, *5*, 723.
36. Bridie, A. L., Bos, J., and Henzberg, S. C. (1973). *Journal of the Institute of Petroleum*, *59*, 263.
37. Ruebelt, C. (1967). *Helgolander wiss Meersunters*, *16*, 306.
38. Przybylski, Z. (1963). *Chemia Analit*, *8*, 601.
39. Reisus, K. (1968). *Fortschr Wasserchem, ihrer, grenzgeb*, *10*, 43.
40. Osipov, V. M. and Belova, T. D. (1968). *Khimiya Tekhnol. Topl. Masel.* *13*, 56.
41. Lurje Yu Yu, (1963). *Nauch. Soobschch; Vses. Nauchno-issled, Inst. Vodosnabzh, Kanaliz Gidrotekhn. Soor zuz henii. i Inzh Gidrogeol.* Ochiska Prom. Stock n Vod, Moscow. 34.

42. Jeltes, R. and Veldink, R. (1967). *Journal of Chromatography*, *27*, 242.
43. Skotnikova, L. A. (1969). *Izy Vyssk Ucheb. Zavod Neft Gas*, *12*, 111.
44. Danyl, F. and Nietsch, N. (1952). *Microchemie, Mikrochem Acta*, *39*, 333.
45. Nietsch, B. (1954). *Angew. Chem.*, *66*, 571.
46. Nietsch, B. (1956). *Gass Wass Warme*, *10*, 66.
47. Nietsch, B. (1956). *Mikrochemica Acta*, 171.
48. Peake, E. and Hodgson, G. W. (1966). *Journal of the America Oil Chemists Society*, *43*, 215.
49. Peake, E. and Hodgson, G. W. (1967). *Journal of the American Oil Chemists Society*, *44*, 696.
50. Golden, J. (1976). *Techniques et Sciences Minicipales*, *71*, 17.
51. Davies, A. W. (1971). *Bulletin of the Belgian Centre for Water Studies 330*, 252.
52. Caruso, S. C. (1967). *Developments in Applied Spectroscopy*, *6*, 323.
53. Kawahara, F. K. (1972). *Journal of Chromatographic Science*, *10*, 629.
54. Kawahara, F. K. (1969). In, *Laboratory Guide for the Identification of Petroleum Products*. 1014 Broadway, Cincinnati, Ohio. US Dept. of the Interior, Federal Water Pollution Central Administration, Analytical Quality Control Laboratory, 41 pp.
55. Jeltes, R. and Van Tonkelaar, W. A. M. (1972). *Water Research*, *6*, 271.
56. Desbaumes, E. and Imhoff, C. (1972). *Water Research*, *6*, 885.
57. Benyon, L. R. (1968). In, *Methods for the Analysis of Oil in Water and Soil*, The Hague, Stickting, CONCAWE, pp. 40.
58. Nadzhafova, K. N. (1970). *Trady Vses. Nanchro-issled. Inst. Vodosnabzh, Kanaliz, Gidrotekhn, Sooruzhenii, Inzh. Gidrogeol, No. 23*, 107.
59. Harva, O. and Somersalo, A. (1958). *Acta Chem. Fenn.* *31*, 384.
60. Simard, R. C., Hasecawa, W., Bondaruk, W., and Headington, C. E. (1951). *Analytical Chemistry*, *23*, 1384.
61. Hellman, H. (1969). *Deutsch gewaesserck Mitt*, *13*, 19.
62. Ruebelt, C. (1967). *Heigolander Wiss Meerseunters*, *16*, 306.
63. Fastabend, W. (1965). *Chemie-Ingr-Tech.*, *37*, 728.
64. Ruebelt, C. (1967). *Gas-u-Wass Fach*, *108*, 893.
65. Ruebelt, C. (1966). *Zeito Analytical Chemistry*, *221*, 299.
66. Osipov, V. M. (1971). *Khimiya Tekhnol Topl Masel*, *17*, 52.
67. Golubeva, M. T. (1966). *Lab Delo*, *11*, 665.
68. Lindgreen, C. G. (1971). *Journal of American Water Works Association*, *49*, 55.
69. Benyon, L. R. (1968). In, *Methods for the Analysis of Oil in Water and Soil*, The Hague, Struchting, CONCAWE, 40 pp.
70. Mallevialle, J. (1974). *Water Research*, *8*. 1071.
71. Martin, P. and Geyer, D. (1974). *Korrespondez Abwasser*, *21*, 202.
72. Shtivel, N. K., Gipokova, L. I., and Smirnova, Z. S. (1985). *Soviet Journal of Water Chemistry and Technology*, *7*, 66.
73. Becconsall, J. K. (1978). *Analyst (London)*, *103*, 1233.
74. Gruenfeild, M. (1973). *Environmental Science and Technology*, *7*, 636.
75. Jeltes, R. (1969). *Water Research*, *3*, 931.
76. Mirzayanov, V. S. and Bugrooyn, F. (1972). *Zavod Lab*, *B8*, 656.

77. Mel'kanovitskaya, C. G. (1972). *Gidrokhim Mater*, *53*, 153.
78. Dudova, M. Ya. and Diterikhas, O. D. (1969). *Didrokhim Mater*, *50*, 115.
79. Matsumoto, G. and Hanya, T. (1980). *Journal of Chromatography*, *194*, 199.
80. Constable, D. J. T., Smith, S. R., and Tomaka, J. (1984). *Environmental Science and Technology*, *18*, 895.
81. Scholz, L. and Altmann, H. J. Z. (1968). *Analytical Chemistry*, *240*, 81.
82. Schnossner, H., Falkenberg, W., and Althaus, H. (1963). *Zeitschrift fur Wasser und Abwasser Forschung*, *16*, 132.
83. Monorca, S., Causey, B. S., and Kirkbright, G. C. (1979). *Water Research*, *13*, 503.
84. Grimmer, G., Dettbarn, G., and Schreider, D. (1981). *Zeitschrift fur Wasser und Abwasser Forshung*, *14*, 100.
85. Borneff, J. and Kunte, H. (1969). *Archives Hygiene Balt*, *153*, 220.
86. Acheson, M. A., Harrison, R. M., Perry, R., and Wellings, R. A. (1976). *Water Research*, *10*, 207.
87. Acheson, H. A., Harrison, R. M., Perry, R., and Welling, R. A. (1976). *Water Research*, *10*, 207.
88. Wedgewood, R. and Cooper, R. C. (1953). *Analyst (London)*, *78* 170.
89. Wedgewood, P. and Cooper, R. C. (1954). *Analyst (London)*, *79*, 163.
90. Wedgewood, P. and Cooper, R. C. (1956). *Analyst (London)*, *81* 42.
91. Hess, G. G., Mckenzie, D. E., and Hughes, B. M. (1986). *Journal of Chromatography*, *366*, 197.
92. Koppe, P. and Muhle, P. (1968). *Vom Wasser*, *35*, 42.
93. Natusch, D. F. S. and Tomkins, B. A. (1978). *Analytical Chemistry*, *50*, 1429.
94. Goldberg, M. C. and Weiner, E. R., (1980). *Analytica Chimica Acta*, *115*, 373.
95. Environmental Protection Agency, (1976). In, *Quality Criteria for Water, Superintendant of Documents*, US Government Printing Office, No 005-09-01049-4, Washington, DC.
96. Semenchenko, L. L. and Kaplin, V. T. (1968). *Zhur Analit Khim*, *23*, 1257.
97. Matsumoto, G., Ishiwateri, O., and Hanya, T. (1977). *Water Research*, *11*, 693.
98. Yorkshire Water Authority, UK (1977). In, *Determination of phenols in potable waters by gas chromatography* Report, Method 655-01.
99. Bengtsson, G. (1985). *Journal of Chromatographic Science*, *23*, 397.
100. Thompson, S. and Eglington, G. (1978). *Geochim, Cosmochim Acta*, *42*, 199.
101. Maktaz, E. D., Batvinova, L. E., and Kruchinina, A. A. (1984). *Soviet Journal of Water Chemistry and Technology*, *6*, 59.
102. Goncharova, I. A. and Khomenko, A. N. (1968). *Didrokhim Mater*, *47*, 161.
103. Johnson, R. W. and Calder, J. A. (1973). *Geochimica Cosmochim Acta*, *37*, 264.
104. Van Hoevan, W., Maxwell, J. R., and Calvin, M. (1969). *Geochim Cosmochimica Acta*, *33*, 877.

105. Mishimura, M. (1977). *Geochim Cosmochimica Acta*, *41*. 1817.
106. Oliver, B. G. and Bothen, K. D. (1980). *Analytical Chemistry*, *52*, 2066.
107. Dietz, F. and Trund, J. (1973). *Vom Vasser*, *41*, 137.
108. Deetman, A. A., Demeuleemeester, P., Garcia, M., and Hanck, G. (1976). *Analytica Chimica Acta*, *82*, 1.
109. Hagenaier, H., Werner, G., and Jager, W. (1982). *Zeitschrift fur Wasser und Abwasser Forshing*, *15*, 195.
110. Bungasser, A. J. and Calarmotolo, J. F. (1977). *Analytical Chemistry*, *49*, 1588.
111. Thompson, J. F., Reid, S. J., and Kantor, E. J. (1977). *Archives of Environmental Contamination and Toxicology*, *6*, 143.
112. Beconsall, J. K. (1978). *Analyst (London)*, *103*, 1233.
113. Haben, H. J., Ching, S. A., Caserette, L. J., and Young, R. A. (1976). *Bulletin of Environmental Contamination and Toxicology*, *15*, 78.
114. Bergtsson, G. (1985). *Journal of Chromatographic Science*, *23*, 397.
115. Bauer, U. (1972). *Gas-u Wasserfach Wasser und Abwasser*, *113*, 58.
116. Fielding, M., McLoughlin, K., and Steel, C. (1977). In, *Water Research Centre Enquiry Report ER 532*, August 1977, Water Research Centre, Stevenage Laboratory, Elder Way, Stevenage, Herts, UK.
117. Reunanen, M. and Kronfeld, R. (1982). *Journal of Chromatographic Science*, *20*, 449.
118. Bush, B., Norang, R. S., and Syrotynstu, S. (1977). *Bulletin of Environmental Contamination and Toxicology*, *18*, 436.
119. Dressman, R. C., Stevens, A. A., Fair, J., and Sinch, B. (1979). *Journal of American Waterworks Association*, *71*, 392.
120. Richard, J. J. and Junk, G. A. (1977). *Journal of American Water Works Association*, *79*, 62.
121. Henderson, J. E., Peyton, G. R., and Glaze, W. H. (1976). A convenivent liquid–liquid extraction method for the determination of halomethanes in water at the parts per billion level. In *Identification and analysis of organic pollutants in water*. (L. H. Keith edition), Ann Arbor Science, Ann Arbor Michigan, *105*, 195.
122. USEPA Part III, Appendix C. Analysis of trihalomethanes in drinking water. *Federal Register*, 44, No 231 68672, 29 November (1979).
123. Method 501.1 (1979). In, *The Analysis of trihalomethanes in finished waters by the purge and trap method*. ESEPA, EMSL, Cincinnati, Ohio, 45628, 6 November.
124. Kirschen, N. A. (1980). *Various Instrument Applications*, *14*, 10.
125. Von Rensberg, J. F. T., Von Higssteon, J. T., and Hassett, A. J. (1978). *Water Research*, *12*, 127.
126. Nicolson, B. C., Bursill, D. B., and Couche, D. J. (1985). *Journal of Chromatography*, *325*, 221.
127. Mieure, J. P. (1977). *Journal of Americal Waterworks Association*, *69*, 62.
128. Department of the Environment (Natural Water Council Standing Committee of Analysts) (1981). In, *Methods for the examination of waters and associated materials chloro and Dromotrihalo Methanes in Water*, HMSO, London.

129. Girenko, D. B., Klisenko, M. A., and Dishchalka, Y. K. (1975). *Hydrobiological Journal*, *11*, 60.
130. Saekmanreva, M., Pal'usova, O., and Szokalay, A. (1977). *Water Research*, *11*, 551.
131. Taylor, R., Bogacka, R., and Krasnicki, K. (1974). *Chemia Analitica*, *19*, 73.
132. Taylor, R. and Bogacka, T. (1969). *Analytical Abstracts*, *17*, 1206.
133. Weil, L. and Ernest, K. E. (1971). *Gas-u-Wassfach*, *112*, 184.
134. Leoni, V. (1971). *Journal of Chromatography*, *62*, 63.
135. American Public Health Association (1978). In, *Standard Methods for the Examination of Water and Wastewater*, 15th edn, Method 509 A. p. 493.
136. American Public Health Association (1978). Method 573, In, *Supplement to the 15th Edition of Standard Methods for the examination of water and wastewater; Selected analytical methods approved and cited by the United States Environmental Protection Agency*, American Public Health Association, American Water Works Association, Water Pollution Control Federation.
137. *Chromopak News* No. 20, June (1979).
138. Kahanovitch, Y. and Lahav, N. (1974). *Environmental Science and Technology*, *8*, 762.
139. Davies, A. W. (1971). *Trib Cebedeau*, *24*, 252.
140. Erney, D. R. (1979). *Analytical Letters*, *12*, 501.
141. Engst, R. and Knoll, R. (1973). *Nahrung*, *17*, 837.
142. Weil, L. and Quentin, K. E. (1974). *Zeit fur Wasser und Abwasser Forschung*, *7*, 147.
143. Laurin, G. (1970). *Bulletin of Environmental Contamination and Toxicology*, *5*, 542.
144. Konrad, J. G., Pionke, H. B., and Chesters, G. (1969). *Analyst (London)*, *94*, 490.
145. Pionke, H. B. (1969). *Analytical Abstracts*, *17*, 2442.
146. Millar, J. D., Thomas, R. E., and Schattenberg, H. I. (1981). *Analytical Chemistry*, *53*, 214.
147. Thompson, J. F., Reid, S. J., and Kantar, K. J. (1977). *Archives of Environmental Contamination and Toxicology*, *6*, 143.
148. Mosinska, K. (1972). *Proceedings of the Institute Przem Org*, *1971*, 253.
149. Beconsall, J. K. (1978). *Analyst (London)*, *103*, 1233.
150. Gorback, S., Haarring, R., Knauf, W., and Werner, H. (1971). *Bulletin of Environmental Contamination and Toxicology*, *6*, 40.
151. Blankit, P. F. (1979). *Journal of Chromatography*, *179*, 123.
152. Tsukioka, T., Talishita, R., and Murakami, T. (1986). *Analyst (London)*, *111*, 145.
153. Shevchuk, I. A., Dubchenko, Y. G., and Naidenova, T. S. (1985). *Soviet Journal of Water Chemistry and Technology*, *7*, 73.
154. Simal, J., Crous Vidal, J., Maria-Charro Arial Boado, M. A., Diaz, R., and Vilas, D. (1971). *An Bromat (Spain)*, *23*, 1.
155. Schafer, M. L., Peeler, J. T., Gardner, W. D., and Campbell, J. E. (1969). *Environmental Science and Technology*, *3*, 1261.

156. Murray, D. A. J., (1975). *Journal of the Fisheries Research Board, Canada*, *132*, 457.
157. Ishikawa, S., Taketami, M., and Shimohara, R. (1985). *Water Research*, *19*, 119.
158. Mallet, V. and Brun, G. L. (1974). *Bulletin of Environmental Contamination and Toxicology*, *12*, 739.
159. Blanchet, P. F. (1979). *ournal of Chromatography*, *179*, 123.
160. Venkataraman, S. and Sathyamurthy, V. (1979). *Journal of the Indian Water Works Association*, *11*, 351.
161. Rice, J. R. and Dishberger, H. J. (1968). *Journal of Agriculture and Food Chemistry*, *16*, 67.
162. Thompson, J. F., Reid, S. J., and Kantor, E. J. (1977). *Archives of Environmental Contamination and Toxicology*, *6*, 143.
163. Zycinski, D. (1971). *Roczn. Panst, Zahl. Hig.*, *22*, 189.
164. Rollo, J. W. and Cortes, A. (1964). *Journal of Gas Chromatography*, *2*, 132.
165. Holden, E. R., Jones, W. M., and Beroza, M. (1969). *Journal of Agriculture and Food Chemistry*, *17*, 56.
166. Cohen, I. C., Norcup, J., Ruzicka, J. H. A., and Wheals, B. R. (1970). *Journal of Chromatography*, *49*, 215.
167. Gutermann, W. H. and Listi, D. J. (1965). *Journal of Agriculture and Food Chemistry*, *3*, 48.
168. Coburn, J. A., Riplay, B. D., and Chan, A. S. Y. (1976). *Journal of the Association of Official Analytical Chemists*, *59*, 188.
169. Lewis, D. L. and Paris, D. F. J. (1974). *Journal of Agriculture and Food Chemistry*, *22*, 148.
170. Handa, S. K. and Dikshit, A. K. (1979). *Analyst (London)*, *104*, 1185.
171. Westlake, W. E., Monika, I., and Gunther, F. A. (1972). *Bulletin of Environmental Contamination and Toxicology*, *8*, 109.
172. Frei, R. W., Lawrence, J. F., and Le Gay, D. S. (1973). *Analyst, (London)*, *98*, 9.
173. Reeves, E. G. and Woodham, D. W. (1974). *Journal of Agriculture and Food Chemistry*, *22*, 76.
174. Frei, R. W., Lawrence, J. F., and Belliveau, P. F. (1971). *Zeit Analytical Chemistry*, *254*, 271.
175. Coburn, J. A., Ripley, B. D., and Chan, A. S. Y. (1976). *Journal of Association of Official Analytical Chemists*, *59*, 188.
176. Fredeen, F. J. L. I., Saha, J. G., and Balba, M. H. (1975). *Pesticides Monitoring Journal*, *8*, 241.
177. Devine, J. N. and Zweig, G. (1969). *Journal of Association of Official Analytical Chemists*, *52*, 187.
178. Bogacka, T. (1971). *Chemical Analysis*, *16*, 59.
179. Bogacka, T. and Taylor, R. (1970). *Chemie Analit.*, *15*, 143.
180. Suffet, I. H. (1973). *Journal of Agriculture and Food Chemistry*, *21*, 591.
181. Hermann, W. D., Tourayre, J. C., and Engli, H. (1979). *Pesticide Monitoring Journal*, *13*, 128.
182. Method S69. (1978). In, *Method for Triazine Pesticides in Water and Wastewater*. Interim pending issuance of methods for organic analysis of

water and wastes. US Environmental Protection Agency, Environmental and Monitoring Support Laboratory, September.

183. Lee, H. B. and Stokker, Y. D. (1986). *Journal of the Association of Official Analytical Chemists*, *69*, 568.
184. Deleu, R., Barthelemy, J. P., and Copin, A. (1977). *Journal of Chromatography*, *134*, 483.
185. McKane, C. E. and Hance, R. J. (1969). *Bulletin of Environmental Contamination and Toxicology*, *4*, 31.
186. McKane, C. E. and Hance, R. J. (1969). *Analytical Abstracts*, *17*, 3849.
187. McKane, C. E., Byast, T. H., and Hance, R. J. (1972). *Analyst (London)*, *97*, 653.
188. Frank, P. A. and Demint, R. J. (1969). *Environmental Science and Technology*, *3*, 69.
189. Payne, W. R., Pope, J. D., and Benner, J. E. (1974). *Journal of Agriculture and Food Chemistry*, *22*, 79.
190. Norris, L. A. and Mongomery, M. L. (1975). *Bulletin Environmental Contamination and Toxicology*, *13*, 1.
191. Richardson, M. L., Webb, K. S., and Gough, T. A. (1980). *Ecotoxical Environmental Safety*, *4*, 207.
192. Fine, D. H., Raunbehler, D. B., Hufffman, F., and Epstein, S. S. (1975). *Bulletin Environmental Contamination and Toxicology*, *14*, 464.
193. Inaba, K. (1987). *International Journal of Environmental Analytical Chemistry*, *31*, 63.
194. Lee, H. B. (1988). *Journal of Association of Official Analytical Chemists*, *71*, 803.
195. Kawano, Y. and Bevenue, A. (1972). *Journal of Chromatography*, *72*, 51.
196. Avery, M. J. and Junk, G. A. (1985). *Analytical Chemistry*, *57*, 790.
197. Wanatabe, N., Yasuda, Y., Katok, Nakamura, T., Funasaka, R., Shimokawa, K., Sato, E., and Ose, Y. (1984). *Science of the Total Environment*, *34*, 169.
198. Uchiyama, M. (1978). *Water Research*, *12*, 299.
199. Waters, J. and Garrigan, J. T. (1983). *Water Research*, *17*, 1549.
200. Yasuhara, A. and Fuwa, F. (1979). *Journal of Chromatography*, *172*, 453.
201. MacDonald, J. C., Bock, C. A., and Slater, G. D. (1987). *Applied Microbiology and Biotechnology*, *25*, 392.
202. Kugemagi, U., Burnard, J., and Terrier, L. C. (1975). *Journal of Agriculture and Food Chemistry*, *23*, 717.
203. Gardiner, J. (1977). *Analyst (London)*, *102*, 120.
204. Gardiner, J. (1975). In, *Technical Memorandum, TM101*, Water Research Centre, Stevenage Laboratory, Herts, UK.
205. Bengtsson, C. (1985). *Journal of Chromatographic Science*, *23*, 397.
206. Annoff, M. and Josefsson, B. (1976). *Analytical Chemistry*, *48*, 1268.
207. Goldberg, M. C., Delong, L., and Sinclair, M. (1973). *Analytical Chemistry*, *45*, 89.
208. Ahnoff, M. and Josefsson, N. (1974). *Analytical Chemistry*, *46*, 658.
209. Ahnoff, M. and Josefsson, N. (1976). *Analytical Chemistry*, *48*, 1268.
210. Wu, C. and Suffet, I. H. (1977). *Analytical Chemistry*, *49*, 231.
211. Fielding, W., Gibson, T. M., James, H. A., McLoughlin, K., and Steep,

C. P. (1981). In, *Organic Micropollutants in Drinking Water*, Technical Report TR 159, Water Research Centre, Medmenham, UK.
212. Ċopin, G. and Barbier, M. (1971). *Oceanography, 23*, 455.
213. Werner, A. E. and Waldichuk, M. (1962). *Analytical Chemistry, 34*, 1674.
214. Zsolnay, A. and Kiel, W. J. (1974). *Journal of Chromatography, 90*, 74.
215. Law, R. J., Marchand, M., Dahlmann, G., and Fileman, T. W. (1987). *Marine Pollution Bulletin, 18*, 486.
216. Eklund, G., Josefsson, B., and Roos, C. (1978). *Journal of High Resolution Chromatography, 1*, 34.
217. Eklund, G., Josefsson, B., and Roos, C. (1977). *Journal of Chromatrography, 142*, 575.
218. Grob, K. and Grob, G. (1970). *Journal of Chromatography, 125*, 471.
219. Grob, K. (1977). *Chromatographie, 10*, 181.
220. Wilson, A. J. and Forester, J. (1974). In, *US Environmental Protection Agency Report, No EPA 600-7-74, 108.*
221. The Oil Companies International Study Group for Conservation of Clean Air and Water, Europe. (1982). In, *Analysis of Trace Substances in Aqueous Effluents from Petroleum Refineries*. Concawe Report No. 6/82.
222. Gomella, C., Belle, J. P., and Auvray, J. (1976). *Techniques et Sciences Municipales, 71*, 439.
223. Hon-Nani, H. and Hanya, T. (1980). *Water Research, 14*, 1251.
224. Hon-Nami, H. and Hanya, T. (1978). *Journal of Chromatography, 161*, 205.
225. Nishi, S. and Horimoto, H. (1968). *Japan Analyst, 17*, 1247.
226. Nishi, S. and Horimoto, H. (1970). *Japan Analyst, 19*, 1646.
227. Pressman, R. C. (1972). *Journal of Chromatographic Science, 10*, 472.
228. Department of the Environment and National Water Council (UK) (1978). In, *Determination of Organic and Inorganic Mercury in Seawater*, HMSO, London, 23 pp.
229. Chau, Y. K., Wong, P. T. S., and Bengert, G. A. (1982). *Analytical Chemistry, 54*, 246.
230. Unger, M. A., MacIntyre, W. C., Greaves, J., and Huggett, R. J. (1986). *Chemosphere, 15*, 461.
231. Meinema, H. A., Burger, W. T., Verslins-Dehaar, G., and Geners, E. C. (1978). *Environmental Science and Technology, 12*, 288.
232. Soderquist, C. J. and Crosby, D. C. (1978). *Analytical Chemistry, 50*, 1435.
233. Hattori, Y., Kobayashi, A., Takemoto, S., Takami, K., Kuge, Y., Sigimae, A., and Nakamoto, N. (1984). *Journal of Chromatography, 315*, 341.
234. Laughlin, R. B., Guard, H. E., and Coleman, W. M. (1986). *Environmental Science and Technology, 20*, 201.
235. Huey, C., Brinkman, F. E., Grim. S., and Iveson, W. P. (1974). In, *Proceedings of the International Conference on the Transport of Persistent Chemical in Aquatic Ecosystems*, eds A. S. W. Freitas, D. J. Kushner, and D. S. U. Quadri, National Research Council of Canada, Ottawa, Canada, pp. II-73—II-78.

236. Chau, Y. K., Wong, P. T. S., and Kramer, O. (1983). *Analytica Chimica Acta*, *146*, 211.
237. Chau, Y. K., Wong, P. T. S., Bengent, G. A., and Kramer, O. (1979). *Analytical Chemistry*, *186*, 51.

6

PURGE AND TRAP AND HEAD-SPACE ANALYSIS TECHNIQUES

6.1 Purge and trap techniques

There are three basic procedures involving equilibriation of the sample with an inert gas suitable for the preconcentration of organics in water samples.

1. We can allow the solution to come to equilibrium with a known volume of gas, and sample this gas (head-space analysis, see Section 6.2).
2. We can extract the volatile materials by pulling a vacuum, on the solution and collect the volatiles either by adsorption or by cold-trapping.
3. We can sweep the volatile materials from the solution by bubbling with an inert gas, again collecting the volatiles by adsorption (Carbon or Tenax GC) or by cold-trapping (purge and trap).

Of the three general methods, the last seems to be the most practical. Theoretically, with high enough concentrations of hydrocarbons the first method, the head-space analysis, should be both the most accurate and the easiest to calibrate. Operationally, it leaves much to be desired, both because of the problems of sensitivity and those of the accommodation of the larger molecules in water. The second method, vacuum degassing, requires much more equipment than the third method and requires that large amounts of water vapour be removed before the sample is injected into the gas chromatograph. The last method is so much less complicated that even with its calibration problems it has been adopted almost universally.

6.1.1 Carbon Adsorption—solvent extraction techniques

Non-saline waters The gas stripping technique is carried out as follows: a large volume of aqueous sample is purged with an inert gas and the effluent containing volatilized organics swept through a trapping medium such as active carbon. The organics are then desorbed from the carbon with a volume of pure solvent which is relatively small compared to the volume of original water sample. Very large preconcentrations of organics in the sample can thereby be achieved. This technique was first described in 1973.[1-3] Piet in 1975[4] described a gas stripping technique for concentrating organic substances in potable water samples and followed this by adsorption on to carbon, desorption from carbon by a solvent and examination of the extract by high resolution gas chromatography in stainless steel capillary columns coupled with an on-line mass spectrometer computer system. The procedure was applied to the identification of micro-organic constituents in potable water before and after ozonization.

Grob and Zurcher[5] have given a detailed description of equipment and procedures for a closed loop method of stripping trace organic constituents from water in order to obtain preconcentration. These workers point out that an important limitation is the volatility range, which includes low and medium molecular weight organic compounds up to about eicosane in the *n*-alkane series. Another limitation is that the most volatile substances, i.e. the substances that are eluted with the solvent used for extraction from the active carbon, cannot be determined due to volatility losses. It is feasible to analyse these substances by direct transfer of adsorbed, material from the charcoal on to the gas chromatographic column, i.e. elimination of the solvent extraction state. One of the parameters that govern the extraction step is the amount of moisture on the charcoal. For several reasons, non-polar or weakly polar solvents with low misibility in water have to be used and moist carbon particles are poorly wetted by such solvents. For this reason, a drying step is used prior to addition of the solvent. No wetability problems occur, even with completely non-polar solvents, when the filter temperature is kept at least 10 °C above the water temperature (30 °C). This temperature difference reduces the relative humidity of the gas passing the filter to about 40 per cent. Extraction can then be started without prior drying and with any solvent.

The higher the extraction efficiency of the solvent, the smaller is the volume of solvent needed for satisfactory extraction. High efficiences are especially important with relatively pure water samples because all available concentration procedures cause severe losses of extracted substances, not just of the most volatile substances. The most efficient solvent, carbon disulphide, unfortunately, presents problems with purification.[6] If a larger volume can be tolerated, methylene chloride, which is easier to purify, is recommended. The extraction of the most volatile substances is surprisingly difficult. The procedure can be applied quantitatively using a system of internal standards. Standard substances that do not occur in the water sample are added so as to give concentrations comparable with those of the substances to be determined, e.g. C_6, C_{10}, C_{14}, and C_{18} 1-chloroalkanes 5.0 μl of solutions in acetone of concentration 1:100 000–1:20 000.

Kuo *et al.*[7] investigated gas stripping and thermal desorption procedures for preconcentrating volatile polar water soluble organics from water samples for analysis by gas chromatography. They investigated the recovery efficiency of the process for many model compounds in aqueous solutions. Detection limits were at the μg litre^{-1} level or lower for most of the compounds studied.

Colenutt and Thorburn[8,9] carried out more work to establish the practical limits of the gas stripping preconcentration technique when applied to standard mixtures and to real water samples containing mg litre^{-1} quantities of organics and to optimize the various factors affecting the efficiency of the technique. Important variables were the flow rate of stripping gas, the time period for which stripping is carried out, the particle size of the adsorbent, and the choice of solvent used for desorption. It was shown that sample contamination is an even more serious problem than low recoveries. Problems of contamination of aqueous samples by adsorption of atmospheric vapours prior to analysis were also considered.

The stripping system used is illustrated diagrammatically in Fig. 7. Nitrogen from a gas cylinder was purified by passing through an active carbon trap and then a flow controller. The gas was then bubbled through the aqueous sample which was contained in a flat bottomed flask. The gas outlet into the sample was through a coarse frit and in all cases the volume of the flask was selected to be as near to the sample volume as possible so reducing any contamination by laboratory air entering the system. After passing through the water the gas was passed through a splash trap to remove any drops of water and then through the active carbon trap.

The active carbon trap was formed from a piece of $\frac{1}{4}$ in o.d. glass

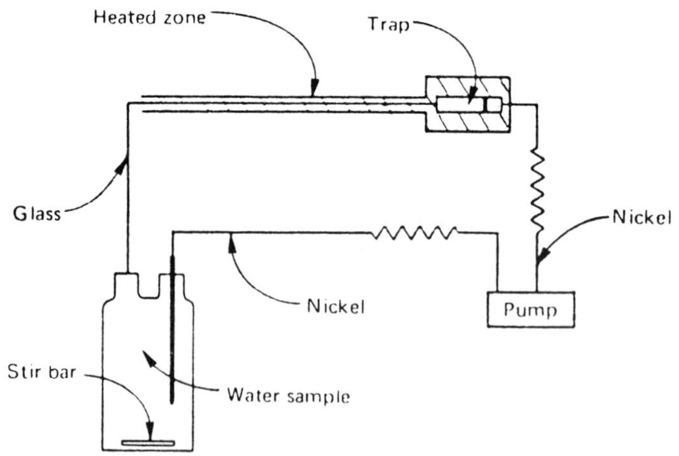

Fig. 7 CLSA flow scheme. From Colenutt and Thorburn[8] with permission.

tube. This was drawn to a fine jet at one end and had an overall length of 8 cm. Active carbon (between 5 and 25 mg) was held in the tube by two pieces of glass fibre yarn. The combined length of the plugs of glass fibre yarn and the active carbon was about 5 cm, leaving an empty portion of tube at the wider end. Flow of gas through the trap was always from the wider end to the narrower end, eliminating any possibility of the adsorbent being blown out of the trap by the gas. The flow control valve ensured constant flow throughout a particular experiment. A soap bubble flowmeter was connected to the outlet end of the trap in order to measure the gas flow at the start and end of each stripping operation. All the glass joints in the system were lined with PTFE sleeves. Connections between glass tubing were made with Swagelok stainless steel fittings incorporating combined PTFE ferrules.

Desorption from the adsorbent was carried out in the following manner, a 5 cm length of PTFE tube was connected to the wider end of the trap. The solvent was then dropped on to the adsorbent by means of a dropping pipette. The active carbon provided a restriction to the flow of the solvent so that when a dropping pipette with a rubber teat was connected to the trap by means of the PTFE tube it was possible either to force the solvent through the adsorbent or suck the solvent up through the solid. Thus by careful manipulation of the pipette bulb the solvent was drawn back and forth over the adsorbent several times before the solvent eventually reached the drawn end of the trap and was collected in a calibrated tube. The volume of solvent used for desorption was variable but was routinely between 50 μl and 1 ml.

Initially the active carbon traps were spiked with known amounts of selected compounds using a microlitre syringe. The percentage recovery of the adsorbate from the adsorbent was measured. Desorption was carried out with a range of solvents in the manner described above. The solutions were then made up to standard volume and analysed by gas chromatography. All the analyses were carried out on a Perkin Elmer Model F11 gas chromatograph fitted with a flame ionization detector (FID) and an electron capture detector (ECD). Column effluent was split in a 1:1 ratio between the two detectors. The columns used for the analysis were either 2 m long, $\frac{1}{4}$ in o.d. glass column packed with 5 per cent OV 17 coated on acid washed Chromosorb W (80–100 mesh) or a 2 m long, $\frac{1}{4}$ in o.d. glass column packed with 5 per cent Carbowax 20 M coated on silanized acid washed Chromosorb W. The oven temperature used varied depending upon the compound being analysed. Nitrogen was the carrier gas at a flow rate of 25 ml min^{-1}.

For, each compound in each solvent the size of the peak on the chromatogram was compared with the peak obtained by injection of a solution of known comparable concentration under the same conditions. It was thus possible to calculate the percentage of the adsorbate

Table 44 Recovery of compounds from 100/120 mesh active carbon traps with various solvents.

Adsorbate	Solvent					
	n-Hexane	Toluene	Carbon disulphide	Diethyl ether	Carbon tetrachloride	Methanol
n-Octane	99	97	99	97	96	90
n-Decane	98	97	99	95	96	90
Toluene	97	–	100	97	96	89
Ethyl benzene	98	99	100	98	96	90
1,2-Dichlorobenzene	96	96	100	95	98	85
Ethanol	90	90	95	92	88	95
n-Butanol	91	88	96	91	92	98
Phenol	94	91	96	92	90	96
Methyl ethyl ketone	91	90	94	93	91	92
Ethyl benzoate	92	91	92	92	90	90
Acetic acid	90	90	95	95	88	92

From Colenutt and Thorburn[9] with permission.

recovered. Typical recoveries of compounds from 100–200 mesh active carbon achieved with particular solvents are listed in Table 44. Recoveries obtained with different particle size adsorbent with carbon disulphide are given in Table 45. A similar pattern of results was obtained using *n*-hexane solvent.

The possibility of contamination from external sources at these exceedingly low concentrations was high. The solutions were prepared in freshly distilled water which had itself been stripped by a flow of nitrogen for 2 h immediately prior to the preparation of the solution. The particle size of the active carbon used in the traps for this work was 100–200 mesh. To ensure that none of the compounds was passing through the traps and failing to be adsorbed in some experiments a second identical trap was connected in series with the first trap. This was analysed at the end of the stripping procedures in the same way as the first trap. In no case was there breakthrough of sufficient quantities to be detected in the chromatographic analysis.

Blank analyses were performed in two ways. If a sample having once been stripped was subjected to a second stripping operation a zero blank was obtained, with the chromatogram showing no peaks. However, if distilled water from the laboratory stock was stripped in the same manner the resulting chromatogram frequently showed traces of some compounds. This was suspected to be as a result of contamination by these compounds present in the laboratory atmosphere. This was

Table 45 Effect of adsorbent particle size on recovery efficiency (percentage of adsorbate recovered after extraction with carbon disulphide).

Adsorbate	Mesh range and corresponding particle size				
	44/60 0.251– 0.353 mm	60/85 0.178– 0.251 mm	85/100 0.150– 0.178 mm	100/120 0.124– 0.150 mm	120/200 0.076– 0.124 mm
n-Octane	92	95	98	99	99
n-Decane	92	96	98	99	99
Toluene	91	95	99	100	100
Ethyl benzene	92	97	99	100	100
1,2-Dichlorobenzene	94	96	99	100	100
Ethanol	86	90	93	95	96
n-Butanol	86	91	92	96	97
Phenol	90	93	95	96	97
Methyl ethyl ketone	86	90	92	94	94
Ethyl benzoate	83	86	90	92	93
Acetic acid	86	90	93	95	95

From Colenutt and Thorburn[9] with permission.

confirmed when it was found that distilled water samples left standing in the laboratory overnight showed higher concentrations of trace pollutants than a sample of freshly distilled water. Problems could arise if water samples with very low concentrations of organic solutes were left exposed in the laboratory since adsorption of further traces of organic compounds could occur. For this reason prior to analysis all samples were kept in securely sealed bottles filled to capacity.

In Table 46 are shown the satisfactory recoveries obtained by the technique for synthetic solutions of polychlorinated biphenyls and chlorinated insecticides.

Boren *et al.*[10] optimized the open stripping system for the analysis of trace organics in water. They examined advantages of the open system compared to the closed loop stripping analysis. Stripping temperatures between 30 and 90 °C were optimized for the recovery of polar compounds, low volatile, semivolatile, and volatile compounds, and phenols. They investigated the effect of extraction solvent (*n*-hexane, benzene, carbon disulphide, methylene dichloride, diethyl ether) on recovery. Extraction efficiency was generally low. A solvent mixture of carbon disulphide/benzene/methanol (63:30:5) was recommended.

Haberer and Schredlseker[11] designed and apparatus for predicting the ease of removal of volatile organics from water by air stripping. The method involves passage of a stream of nitrogen through 2 litres of the aqueous liquid for measured intervals of time, and adsorption of the volatilized organic molecules on to activated carbon. The residual organic content was determined by gas chromatography. The method can be used to determine the rate constant for the stripping effect, from

Table 46 Recovery of pesticides and polychlorinated biphenyls by stripping.

Solute	*n*-Pentane	Carbon disulphide	Carbon tetrachloride	Diethyl ether
Aldrin	95	96	96	88
Dieldrin	92	94	95	85
Lindane	94	94	95	80
p,p'-DDT	95	92	95	85
Endrin	90	92	95	85
Heptachlor	92	94	98	90
Methoxychlor	92	90	94	88
Aroclor 1221	94	92	92	90
Arolor 1248	96	95	94	89
Aroclor 1254	96	98	96	92

From Colenutt and Thorburn[9] with permission.

which the time required to reduce the concentration by half (or any other desired amount) can be calculated.

Closed loop stripping analysis is a further refinement of the technique in which preconcentration factors of up to one-million-fold are possible. Westendorf[12] has described a commercially available apparatus, the Tekman CLS-1 instrument, for the semi-automatic determination of traces of organics in water. This instrument can accommodate any sample size up to 4 litres. Analysis of fuels in water by closed loop stripping analysis can provide an excellent fingerprint of the sample. The detection limit obtainable by closed loop stripping analysis is even less than 1 μg litre^{-1} which is 3–4 orders of magnitude more sensitive than any other technique. Detection limits for other fuels, such as gasoline, kerosene, or jet fuel are even lower.

Concawe[13] have applied the gas stripping analysis technique developed by the Environmental Protection Agency[14] for the preconcentration of organics in oil refinery wastes prior to their identification and determination by gas chromatography—mass spectrometry. The method has been used to determine down to 1 μg litre^{-1} of benzene, toluene, ethylbenzene, and mono-and di-chlorobenzenes in aqueous refinery wastes[15] and up to 29 purgeable chlorine and bromine containing organics in aqueous industrial wastes. For the determination of purgeable trace aromatic aliphatic hydrocarbons a flame ionization detector should be used. For the determination of purgeable halogenated organics a halide selective detector should be used (the Hall 700 A electrolytic conductivity detector is recommended, but the alkali flame ionization or electron capture devices could be used).

Graydon *et al.*[16] sought to combine the advantages of closed loop stripping analysis using activated carbon filters with those of thermal desorption and have evaluated the operating parameters for quantitative applications. The advantages of closed loop stripping analysis include a large enrichment factor, minimal contamination, and closed circuit recycling of substances that break through the carbon filter. One of the limitations is an inability to determine compounds that elute under the gas chromatographic peak of the solvent (usually carbon disulphide). To remedy this, thermal decomposition closed loop stripping analysis was tested in the laboratory. Sixteen volatile compounds with boiling points in the range − 30 to 120 °C were determined with recoveries of 12–52 per cent in good agreement with the results of other methods. The filters were desorbed by direct insertion into the injection port of the gas chromatograph. The method is applicable to highly contaminated waste waters but care must be taken to ensure that the adsorbent is not overloaded.

Coleman *et al.*[17] optimized purging efficiency and quantification of organic contaminants from water using a 1 litre closed loop stripping apparatus and computerized capillary column gas chromatography—mass spectrometry. These workers studied four ways of improving the recovery and precision of the 1 gallon closed loop stripping apparatus: by purging a smaller volume; by better control of the temperature of the gas entering the filter and of the filter itself; by increasing the temperature of the sample water bath; and by eliminating as much metal as possible in the closed loop. The 1 litre system was developed and optimized for maximal purging efficiency of 1-chloroalkane standards spiked in water. The optimal conditions for efficient recovery were established and more than 95 per cent recovery was achieved for several chlorinated paraffins. Recovery studies of a 22 compound standard mixture of organics, using two 1-litre closed loop stripping devices of different design, showed that both systems had similar recovery efficiences, about twice as good as those for 1-gallon devices. Relative standard deviations showed similar improvement.

Blanchard and Hardy[18] collected volatile compounds which had been nitrogen purged from water by permeation through a silicone polycarbonate membrane and adsorption on to charcoal contained within the sampling device. Analysis was performed by desorption with carbon disulphide followed by separation and quantification by capillary gas chromatography. The method provided time-weighted average concentrations not possible with the purge and trap method. A linear relationship existed between the amount of volatile organic component collected and the product of sampler exposure time and component concentration in the environment. Results showed that the method compared favourably with the purge and trap method for precision and accuracy. Advantages of the method included the inexpensive nature of the sampling device, its simplicity, that fact that it could be reused over extended periods of time and that no power supply was required for sampling.

Melcher and Caldecourt[19] described a gas chromatographic injection and concentration technique, based on the delayed injection principle for the direct detection of trace organic compounds in water. The organic samples are retained for extended periods on a collection precolumn while the air or water is vented and then injected into an analytical column with application of a high electric current (the heat source) across the precolumn. The optimal desorption temperature is 150 °C and the optimal heating rate 20 °C s^{-1}. Application of the technique for the determination of glycol ether mixtures, phenol, and soluble aromatic compounds in water is described. The technique is considered particularly useful for determination of compounds which are difficult to extract or purge from water. Most of the published work discusses the use of active carbon for the adsorption of organics released from the water sample upon gas purging. However, some work has been carried out using alternate adsorbents,[20-22] (Section 6.1.2).

Otson and Chau[23] evaluated a purge and trap technique for the preconcentration and determination of μg litre^{-1} levels of 51 target organics in water samples. The 51 compounds from the US EPA purgeable priority pollutants list were each spiked at 1, 10, and 50 μg litre^{-1} into purified water. Detection limits were below 1 μg litre^{-1}, recoveries were greater than 70 per cent, and precision for recoveries over all three concentrations were less than 20 per cent RDS. Improved detection limits were obtained for several water soluble and halogenated compounds when concentrator trap composition was altered, transfer line temperature decreased, and sparger vessel temperature increased. The effects of sample transport and storage on the levels of the organics in spiked water samples were investigated.

Lopex-Avila *et al.*[24] studied the applicability of a purge and trap technique and the use of a photoionization detector and a Hall electrolytic conductivity detector connected in series for determining US EPA SW-846 Methods 8010 and 8020 compounds. Method precision was determined using ground water samples spiked with selected test compounds. The highest relative standard deviation value was 63 per cent for chloromethane. Other values ranged from 1.5 to 30 per cent with a mean value of 12 per cent. Average recoveries ranged from 82 to 104 per cent for six determinations. Method detection limits were determined using the SP-1000 column and the two detectors connected in series. Method detection limits for water samples ranged from 0.1–1 μg litre^{-1}. Lower detection limits might be achieved by using larger samples. Investigations with wide-bore open tubular capillary columns showed good results for the VOCOL or DB-624 columns for the primary analysis, although neither column could completely resolve all compounds.

Lopez-Avila *et al.*[25] also examined the feasibility of using the purge and trap technique, with a wide-bore column directly connected to a mass spectrometer without an open slit interface for the analysis of 34 volatile compounds listed in the US EPA priority pollutant list. Two wide-bore capillary columns were evaluated, VOCOL and 007, neither of which was able to resolve all 34 compounds completely. However, the analysis was faster than for a packed column and resolutions of 20 compounds at 50 μg litre^{-1} per component were achieved with the VOCOL column (diphenyldimethyl polysiloxane with cross-linking moieties) and of 16 compounds with the 007 column (95 per cent methyl, 5 per cent phenyl silicone). The VOCOL column was further tested for performance with nine ground water samples containing unknown concentrations of organics and spiked with three internal standards. The instrument response was linear from 20 to 200 μg litre^{-1} for most compounds.

Mosesman *et al.*[26] investigated factors influencing the analysis of volatile pollutants by wide-bore capillary chromatography and a purge and trap system using five volatile gas mixtures: dichlorodifluoromethane, chloromethane, vinyl chloride, bromomethane, and chloroethane. The factors studied were initial column temperature (10 or 35 °C), carrier gas (helium), flow rate (5, 10, 15 ml min^{-1}), speed with which pollutants were desorbed from the trap, and the type of detector used (flame ionization detector or electrolytic conductivity detector). After optimizing the analytical conditions for the gases, a mixture of 36 volatile pollutants was analysed. The remaining 30 compounds were refocused at the column inlet, producing sharp, well-resolved peaks. A carrier gas flow rate of 10 ml min^{-1} and an initial column temperature of 10 °C were the optimal conditions for purge and trap analysis of volatile priority pollutants from a VOCOL wide-bore capillary column. A new experimental trap improved the chromatography of the volatile pollutants. The choice of detectors for the analysis was critical.

Duffy *et al.*[27] used a purge and trap technique to preconcentrate chlorinated hydrocarbons, benzene, and toluene in potable water and industrial effluents. The purge and trap device was connected to a capillary column gas chromatograph equipped with electrolytic conductivity and photoionization detectors. Chichester-Constable *et al.*[28] have described an improved sparger unit for the preconcentration of organics from landfill water samples containing a high proportion of solids. Volatile organic solvents in the 1–100 μg $litre^{-1}$ range could be determined. Some further applications of the purge and trap technique with carbon adsorption are listed in Table 47.

Sea-water Whilst much of the literature on this subject is concerned with non-saline water samples, it is believed that many of these procedures will also work satisfactorily with sea-water; indeed, the presence of salts in the sample may assist in the removal of volatiles.

In the collection of organic materials, separation of the more volatile materials can be achieved by transferring them into the vapour phase for collection and concentration. The usual method for effecting such a transfer is to bubble some inert gas through the liquid. The effluent gas is then passed through an adsorbent such as carbon to collect the organic materials.[40,41] The solvent desorbed material is then analysed by gas chromatography or by linked gas chromatography–mass spectroscopy.

The Oil Companies International Study Group for Conservation of Clean Air and Water, Europe (Concawe) has made a detailed study[42] of the application of gas stripping to the determination of hydrocarbons in amounts down to parts per billion in water. In this procedure the water sample is purged with nitrogen and helium and the volatiles trapped on a solid adsorbent such as activated carbon. The organics are then released from the carbon by heating and purging directly into a gas chromatograph linked to a mass spectrometer. For chlorine and bromine

Table 47 Applications of purge and trap technique (carbon adsorption–solvent extraction) to natural waters.

Compound	Desorption solvent	Analytical finish	Sensitivity	Ref.
Alkanes	CS_2	GLC	1 μg litre^{-1}	29–32
Chloroalkanes	CS_2	GLC	1 μg litre^{-1}	31
Aromatics	CS_2	GLC	–	32–33
Alicylics	CS_2	GLC	–	32
Dioxane, 2-butanane, 4-methyl-2-pentanone, butoxylethanol	CS_2/CH_3OH	GLC	1 μg litre^{-1} (4 litre sample)	34
Vinyl chloride	CS_2	GLC	–	35
Chloroaliphatics	CS_2	GLC	–	36
Monochlorobenzene, monochlorotoluene	CS_2	GLC	–	37
Acetic acid	–	GLC	0.02 M	38
1,1-Dichloroethane	–	GLC	–	38
Odour components (2-methylisoborneol, 2,3,6-trichloroanisole, geosmin	–	GLC	–	39

containing impurities a halide selective detector such as the Hall electrolytic conductivity detector is used on the gas chromatograph. Alternatively, an alkali flame ionization detector or an electron capture detector could be used.

Colenutt and Thorburn[43,44] have also described the procedure using gas stripping of the aqueous sample followed by adsorption to active carbon from which surface they are taken up in an organic solvent for gas chromatographic analysis. They optimized conditions for the determination of parts per billion of pesticides and polychlorinated biphenyls.

The closed loop gas stripping system has been discussed by various workers.[45-49] In this technique organic compounds are removed from water by purging with a gas saturated with water vapour (Fig. 7). Volatile and semi-volatile compounds will partition out into the head space and are swept to an activated charcoal trap. The charcoal will retain organics while allowing the purge gas to pass through. The purge gas is then returned to repurge the sample via a pump. At the end of the purge time, typically 2 h, the trap is removed and fitted with a glass collection vial. Organic compounds are extracted from the charcoal with a small volume of a suitable solvent such as carbon disulphide or dichloromethane which is then collected and injected into a capillary gas chromatograph or a capillary gas chromatograph coupled with a mass spectrometer.

Grob and Zurcher[45-48] have carried out very detailed and systematic studies of the closed loop gas stripping procedure and applied it to the determination of parts per billion of 1-chloroalkanes in water. Westerdorf[49] applied the technique to chlorinated organics and aromatic and aliphatic hydrocarbons. Waggott and Reid[50] reported that a factor of major concern in adapting the technique to more polluted samples is the capacity of the carbon filter which usually contains only 1.5–2 mg carbon. They showed that the absolute capacity of such a filter for a homologous series of 1-chloro-*n*-alkanes was 6 μg for complete recovery. Maximum recovery was dependent on carbon number being at a maximum between C_8 and C_{12} for the 1-chloro-*n*-alkane series. It is important, therefore, to balance the amount of sample stripped with the capacity of the carbon filter to obtain better than 90 per cent recoveries.

6.1.2 Tenax GC or Chromosorb Adsorption—Thermal Desorbtion

Non-saline waters Dowty *et al.*[21,22] used a modified head space technique involving gas purging with subsequent adsorption of organics on poly(*p*-2,6-diphenylphenylene) oxide adsorbent (Tenax GC) for the monitoring of low μg litre^{-1} amounts of low molecular weight organics in potable water in very low concentrations. In this method volatile organics were thermally desorbed from the water sample and displaced into the head space by purging the sample with ultrapure helium. These displaced organics were subsequently adsorbed on to Tenax GC. A glass liner which fits into a modified injection port of the gas chromatograph was packed with this adsorbent. The liner was then inserted into the injection port where the organics were thermally desorbed. These organics were then swept on to the precolumn of a gas chromatograph maintained at dry ice/methanol temperatures, where they condensed in order to ensure a 'plug' injection on to the chromatographic column. After condensation in this manner for a fixed period of time, the temperature of the precolumn was quickly raised and the organics were 'flashed' on to the chromatographic column (91 m × 0.05 cm stainless steel capillary coated with 10 per cent SF-96 and 1 per cent Igepal -CO-880) where resolution of the components was achieved. The procedure was automated. Figure 8 illustrates a chromatogram of organics obtained from 1 litre of New Orleans potable water using this procedure.

It is possible to also collect the volatiles in a cold trap[51] (see Section 6.1.3). A more favoured technique is the collection of the gases by adsorption on some support such as one of the Chromosorbs, or Tenax GC.[51,52] The volatiles are then desorbed by heating and injected into a gas chromatograph.

Hammers and Bosman[53] quantitatively evaluated a simple dynamic technique for non-polar pollutants in aqueous samples at the μg litre^{-1} level. The aqueous sample was purged with nitrogen and *n*-alkanes, benzene, and alkylbenzenes collected in a Tenax trap. The trap tube was mounted at the injection port of a gas chromatograph, where analytes were trapped at the head of the capillary column by cooling with liquid nitrogen. Quantitative date were obtained using flame ionization detection. Distilled water, natural ditch water, and a soil water suspension, each spiked with known amounts of solutes, were analysed. Recoveries of alkylbenzenes were higher than expected and might have been caused

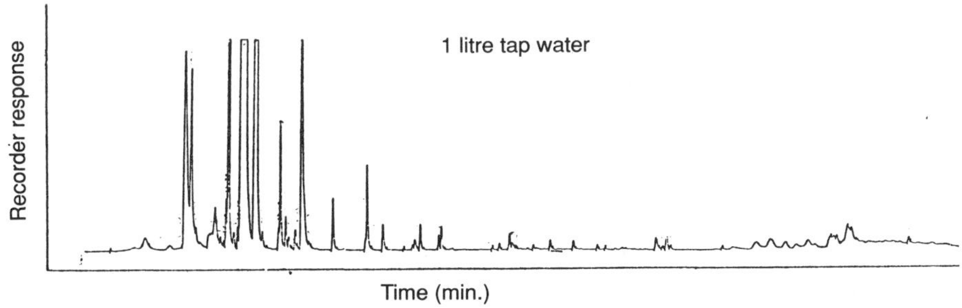

Fig. 8 Gas chromatogram of volatile organics present in one litre of potable water. Resolution achieved on a 91.0 m × 0.05 cm stainless steel capillary column coated with 10 per cent SF-96 and 1 per cent Igepal CO-880. Chromatographic conditions described in the text. From Dowty *et al.*[21] with permission.

by transfer of solute molecules to the gas above the water sample when they were adsorbed to the gas–water interface of the gas bubbles. Quantitative recoveries were obtained for alkanes and alkylbenzenes from distilled water and ditch water but not for chlorinated ethenes. The applicability of head-space sampling under these conditions was best restricted to analytes with Henry's coefficients similar to that of benzene or larger. Detection limits for alkanes and alkylbenzenes and for the examined chlorinated ethenes were 1 and 5 μg litre^{-1}, respectively, The use of a number of strip vessels and traps would reduce analysis time to less than the 2 h.

Haloforms have been preconcentrated by adsorption on Tenax GC followed by thermal desorption on to a gas chromatograph[54,58–63] or an atmospheric pressure microwave emission detector.[55–57] Vinyl chloride has been preconcentrated by adsorption on to silica gel followed by thermal desorption and analysis by gas chromatography using a halogen specific detector.[64]

Sea-water Perras[65] has pointed out that when the concentration of hydrocarbons in sea-water is great enough, as, perhaps, after a petroleum spill, the emergent gas stream can be sampled directly. This is seldom the case in true oceanic samples, however, and some form of concentration is needed.

It is possible to collect the volatiles in a cold trap.[66] A more favoured technique is the collection of the gases by adsorption on some support such as one of the Chromosorbs, or Tenax GC.[66–68] The volatiles are then desorbed by heating and injected into a gas chromatograph. In a gas chromatography—mass spectrometry analysis of a sample of volatile materials concentrated from inshore waters on to Tenax GC, MacKinnon[69] reported tentative identification of methyl isopropyl ketone, bromoform, 4-methyl-2-pentanone, 2-hexanone, and 2-hexanal.

6.1.3 Cryogenic trapping

Non-saline waters Rayer and Hennequin[70] described a purge and trap method for the separation of volatile organic compounds from the aqueous phase. Samples are degassed into the trap at −70 °C for 15 min under a flow of nitrogen; the trap is then closed by metal discs and stored at 4 °C before the gases are desorbed at 40–45 °C directly into the injection port for gas chromatographic—mass spectrometric analysis. The system was used to determine about 15 compounds, in the concentration range 0.05–10 μg litre^{-1} in a wide range of waters.

Bodings *et al.*[71] showed that the temperature of the first cold trap was important and high concentrations of volatile compounds gave irregularities. The efficiency of different types of cryogenic trap depended on purge flow rate, total purge volume, and temperature of trapping. Hydrocarbons have been preconcentrated by cryogenic trapping of volatiles released by purging the sample with an inert gas prior to examination of the trappings by gas chromatography[72] or gas chromatography–mass spectrometry.[73] Down to 1 μg litre^{-1} hydrocarbons could be determined by this procedure.

Smith[74] has discussed the design and features of an on-line purge and cold-trap preconcentration device for the rapid analysis of volatile organic compounds in aqueous samples. A theoretical model describing ice formation and growth in the capillary trap was evaluated. Synthetic mixtures covering concentrations in the range 0.1–10 μg litre^{-1} and different chemical classes were used to study the effects of various process factors on the efficiency and selectivity of water removal as well as on the purging recovery. Either a condenser or a Nafion water permeable membrane could be used to remove water selectively from non-polar compounds. Polar and high boiling non-polar solutes were partly or completely lost by both methods. The importance of the concentration of solutes, flow rate relative to the volume of the purge gas, and the temperatures of the condenser, cold trap, and sample was emphasized. Non-polar volatile organic compounds could be enriched to give a detection limit of 1 μg litre^{-1}. Process time could be reduced by purging at elevated sample temperatures and/or at large purge gas flow rates.

Cochrane[75] also used a Nafion tube dryer in a purge-whole column cryotrapping method for selectively removing water from the analyte containing purge stream during the gas chromatography analysis of volatile aromatics in aqueous samples. The addition of the tube drier meant that water transferred to the column during sample purge could neither extinguish the flame of the flame ionization detector nor distort chromatographic peak shape. The method was applied to the analysis of many volatile chlorinated and fluorinated hydrocarbons and purgeable aromatics found in contaminated aqueous samples. Gryder-Boutet and Kennish[76] described equipment for the integration of an automated head-space sampler, with whole column cryotrapping and mass selective detection. Cryogenic trapping (−80 °C) gave excellent reproducibility in the μg litre^{-1} range and increased sensitivity to the litre^{-1} range. Coated fused silica capillary columns were used for gas chromatographic separation, and mass spectral data were obtained by selective ion monitoring. This method was recommended as an alternative to the US EPA method 624 which used purge and trap technology with packed columns.

The above methods all employ direct cryogenic trapping at −70 to −80 °C of volatiles swept from the samples. An alternate method is to sweep the sample volatiles into a trap containing Tenax GC adsorbent immersed in liquid nitrogen. Cailleux *et al.*[77] preconcentrated trihalomethanes, benzene, and toluene from natural waters by bubbling inert gas through the liquid or by passing the gas over the surface, followed by trapping on a polymer absorbent cooled by liquid nitrogen. For injection, the trap was simultaneously heated and purged with carrier gas. Detection limits of approximately 0.5 μg trihalomethane per litre for 20 ml sample volumes were achieved with a relative standard deviation of 20 per cent.

Henatsch and Juttner[78] used cryoadsorption to preconcentrate low boiling organic sulphur compounds from anoxic lake water samples. The method was based on a stripping system with a cryotrep containing odour trapping Tenax TA cooled with liquid nitrogen. Heat desorption was used to transfer the compounds from cryotrap on to UCON-coated glass capillary gas chromatography columns without modifying the injection system. Sulphur compounds were easily detected by a variety of detectors including flame photometric or mass-selective detectors. Ethanethiol, carbon disulphide and dimethyl sulphide were found in anoxic fresh-water lake samples. Hydrogen sulphide interferences were overcome by treatment with iron(III) chloride and excessive amounts of methane were removed by using solid carbon dioxide instead of liquid nitrogen for cryotrap cooling.

Sea-water Material removed from sea-water by stripping in this manner can be concentrated by trapping in a loop immersed in a cooling bath. The usual cooling baths are liquid nitrogen or solid carbon dioxide with or without an organic solvent. The major disadvantage of the liquid nitrogen bath, along with the cost and the limited availability, particularly aboard ship, is that it condenses carbon dioxide and water, along with the organic materials actually desired. These trapping techniques have been used by many workers and are usually described in conjunction with a gas chromatographic determination of some fraction of the organic materials. An example of this kind of separation is the work of Novak *et al.*[79] They improved the yield of volatile organics by salting out the volatiles with sodium sulphate.

Vacuum removal of volatiles has also been employed in the preconcentration of organics from sea-water. This system, involving the removal of volatiles under vacuum can be set up in two ways; either as a flow-through or as a batch process. As a flow-through process, the sample is drawn continuously through the system, and the gases taken off by the vacuum pass through a sampling loop. Periodically, the material in the loop is injected into the gas chromatograph. In this manner it is possible to derive almost continuous profiles of volatile hydrocarbon concentrations.[80] In the batch mode, a larger sample can be treated over a longer period, and the volatiles collected by cold-trapping or adsorption. These techniques are not as fast as flow-through sampling, nor do they permit semi-continuous profiling, but they result in a greater concentration of the hydrocarbons, and thus in greater sensitivity.[81] Cerwen[82] used the method for the analysis of ketones and aldehydes (acetone, butyraldehyde, and 2-butanone) in sea-water. Halocarbons were similarly separated from environmental samples by Kaiser and Oliver.[83]

6.2 Head-space analysis

6.2.1 Non-saline waters

This technique differs from purge and trap techniques in the following respect. In purge and trap, the water sample is purged with a gas and the organics removed and adsorbed on to a solid, usually active carbon. The adsorbed organics are then desorbed in a small volume of an organic solvent, usually chloroform or carbon disulphide which is subsequently analysed by gas chromatography. Although capable of achieving very high concentration factors a disadvantage of this technique is that the organic solvent and impurities therein can cause difficulties at the gas chromatographic state of the analysis. Head-space analysis is solventless in that the aqueous sample is heated say to 30 °C in the presence, usually, of inorganic salts which help to drive the organics into the inert gas filled head-space above the sample. Portions of the head-space gas are then withdrawn by syringe for gas chromatographic analysis. Thus head-space analysis unlike purge and trap, is a static method. By attention to detail in the design of the appratus and operating conditions, it is possible by this technique to obtain concentration factors of up to 100 for the organics in the original water sample.

Wyllie[84] has observed that purge and trap methods are less readily automated, liable to lose of light volatiles, readily contaminated by samples containing a very high concentration of some volatiles, subject to contamination by the ambient air, expensive on glassware and/or glassware washing, and have a low sample throughput. Head-space was more readily automated, usable on other matrices besides drinking water, portable (and hence available for field use), and labour-saving to the extent that all glassware was disposable; but it too was liable to contamination by the ambient air. Under suitable conditions, the techniques were equally sensitive, but head-space had the advantage in precision.

Munz and Roberts[85] elucidated the interactions of Henry's constant, solubility, and activity coefficients by the multiple equilibration method of a closed system with analysis of the aqueous phase. Methanol and isopropanol were used as model cosolvents in a study of eight common chlorinated solvents. One-hundred-ml gas tight syringes were used as containers. Solvents were spiked with solute and equilibrated for 25 min before expelling the gas phase and analysing the liquid phase by GC. Experimental results were compared with predictions from a semi-empirical thermodynamic model and from vapour pressure and solubility date (UNIFAC). No effect of solute concentration on the solute's Henry's constant was found up to solute–liquid mole fractions of 0.001. Very high cosolvent concentrations in excess of 10 g litre^{-1} were required to reduce the Henry's constant. No effect was observed in multisolute systems at low concentrations. The Henry's constant of the organic compounds increased by a factor of approximately 1.6 for each 10 °C rise in temperature and was considered to be the most important factor in determining the removal of volatile solutes in gas–liquid contacting systems. The UNIFAC model predicted relative effects of concentration and cosolvent but could only be applied with caution to predict temperature effects.

Friant and Suffet[86] obtained 66-fold enrichment factors by:

(1) increasing the temperature of the water sample from 30 to 50 °C;
(2) saturating the water sample with sodium sulphate; and
(3) adjusting the pH of the original sample to 7.1.

Their method used a Tracor MT-550 gas chromatograph equipped with dual flame ionization detectors or equivalent, and a chromatographic column 8 ft × 1/8 in i.d. stainless steel packed with 20 per cent SE-30 on 80/100 mesh Gas Chrom Q. The column temperature was maintained isothermally at 130 °C. Inlet, outlet, and detector temperatures were 180, 200, and 200 °C respectively. Glassware in contact with both water and vapour phases was silanized to minimize surface adsorption. First the glassware was washed with detergent. This was followed by rinsing with distilled water and air drying. The dry surface was treated with Glass-treet (Alltech Associates) and rinsed with anhydrous methanol.

The sampling bottle used for laboratory test conditions was a modified 2 litre Pyrex reagent bottle. The neck was reformed by using 30 glass O-ring joints and the top was rounded and sealed. The two joints were sealed by a Buna rubber O-ring and a compression clamp. To allow syringe sampling of the vapour two ($\frac{1}{4}$ in diameter) glass side arms were attached. A stainless steel ($\frac{1}{4}$ to $\frac{1}{8}$ in) Swagelok reducing union was attached to the glass sidearms by $\frac{1}{4}$ inch rubber O-rings and a Swegelok nut. The sampling port was sealed by a chromatographic silicon septum. The gas syringe was a Precision Scientific (Baton Rouge, LA) pressurizable Series A-2 10.0 ml gas syringe equipped with sideport needle and stop/go valve. The valve permitted the sampling of larger vapour volumes by allowing compression of the sample prior to injection, thereby reducing peak broadening. The syringe also permit easier injections since the pressure in the syringe can be equal to the GLC column head pressure.

All experiments were run isothermally in a constant temperature air bath controlled to ± 0.5 °C. All samples were stirred on a magnetic stirrer. Anhydrous sodium sulphate 475 g litre^{-1} (3.35 M) when dissolved in 0.2 M orthophosphate buffer) was placed in a clean sample bottle at the required experimental temperature 16 h prior to the analysis to reduce the time required to dissolve the salt. A sampling temperature of 50 °C, was used. Vapour samples were withdrawn through the sidearm sampling port of the sample bottle with the gas syringe. The syringe was flushed with vapour phase prior to withdrawing the sample. After sampling the vapour, the syringe valve was closed and the volume reduced to $\frac{1}{10}$ of the sample volume followed by direct injection into the gas chromatographic column. The general experimental conditions were:

Equilibration time	5 h
Vapour volume	1100–1200 ml
Liquid volume	1000–1040 ml
Volume of vapour samples	5.0 ml
pH	7.1
Sodium sulphate concentration	0.00–3.35 M
Ionic strength	0.00–10.02
Orthophosphate buffer	0.20 M

Table 48 is a three-dimensional presentation of the results obtained from the 3 × 3 × 2 statistical experiments. Methyl ethyl ketone data are shown as an example. It is seen that the highest partition coefficient (*K*) for methyl ethyl ketone is obtained with 3.35 M sodium sulphate, pH 7.1,

Table 48 Three-dimensional presentation of the experimental results for the analysis of variance of methyl ethyl ketone, $K \times 10^{-3}$.

	Temperature (°C)					
	30			50		
	Salt concentration (M)					
	0.00	1.41	3.35	0.00	1.41	3.35
pH 4.5	4.19	21.3	118	21.3	39.8	234
pH 7.1	3.90	20.0	109	20.0	37.6	260
pH 9.1	4.56	18.7	105	18.7	35.0	229

From Friant and Suffet[86] with permission.

Table 49 Partition coefficient of model compounds at the optimum and reference states.

Compound	$K \times 10^{-3}$	
	Optimum conditions, pH 7.1, 50°C 3.35 M[a]	Reference condition, pH 7.1, 30°C, no salt[a]
Methyl ethyl ketone	260	3.90
Nitroethane	72.5	2.89
Butanol	44.3	0.746
Dioxane	13.7	0.278

[a] 0.2 Morthophosphate buffer.
From Friant and Suffet[86] with permission.

and 50 °C. In Table 49 are shown partition coefficients for four different organic compounds under these conditions, relative to conditions where sodium sulphate is not added and the temperature is reduced to 30 °C.

The calculated theoretical detection limits for the four organic compounds examined by Friant and Suffet[86] are shown plotted against the reciprocal of the partition coefficient (*k/K*) in Fig. 9. The theoretical limits were determined by assuming a nominal flame ionization detector limit of 50 ng absolute amount injected. The detection limits ranged from 50 μg litre^{-1} for methyl ethyl ketone to 740 μg litre^{-1} for dioxane. The smaller *I/K* is the lower the detection limit. Figure 9 can be used

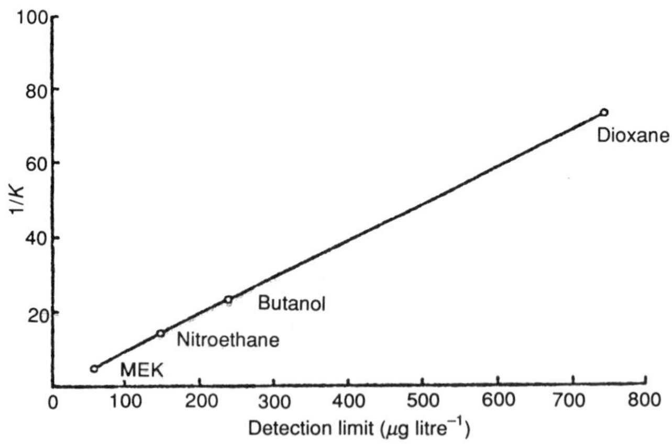

Fig. 9 Theoretical detection limits for model compounds plotted versus the reciprocal of the partition coefficients; the nominal FID detection limit is 50 ng and the amount is 5.0 ml of injected vapour. From Friant and Suffet[86] with permission.

to determine the theoretical detection limits for any solute from the measured partition coefficient under comparable conditions.

The general 'screening' procedure for quantitative analysis with semi-quantitative evaluation has been utilized for taste and odour profiling of water. A primary reason for the investigation of head-space analysis was to subsequently develop a method that would isolate volatile organics causing taste and odour in potable water. Isolation of possible taste and odour organics under the enhanced conditions allows determination of the initial water phase concentrations. These concentrations can then be used to determine the vapour phase concentrations of the organic compounds presented to the consumer of drinking water. All that is needed is the partition coefficients of the organic compound at the enhanced and potable water conditions, respectively.

Application of the method of head-space analysis as described was utilized for gas chromatography—mass spectrometry of a potable water sample in Philadelphia. The results of mass spectral identification of the compounds found in the potable water showed the presence of toluene, two C-2 benzene isomers, chloroform, chlorodibromomethane, dichlorobromoethane, and 1,1,2,2-tetrachloroethane. The analysis was completed in 125 ml bottle containing 100 ml of potable water. A 50 μl gas sample was injected on to the gas chromatograph using a ^{63}Ni electron capture detector.

Malten and Vreden[87] applied the gas chromatographic head-space analysis technique to the detection of organic volatiles in very small volumes of aqueous samples. Cowen *et al.*[88] have described the construction of a septumless injection device for delivering head-space gases into a gas chromatograph.

Distillation techniques have been combined with head-space analysis as a means of improving preconcentration factors of up to 100.[89-91] The distillation technique has been used[89,91] to concentrate volatile polar water-soluble organic compounds in the distillate for later head-space gas chromatographic analysis. A fast and reliable head-space gas injection system equipped with an evacuated gas sampling valve was used. Methanol, ethanol, acetone, 2-propanol, and methyl ethyl ketone and other volatile polar organics can be determined at concentrations ranging from 0.008 to 13 mg litre^{-1}. Combining the head-space gas chromatographic method of analysis with sample preconcentration by the distillation technique made it possible to determine volatile polar organics at the μg litre^{-1} level.

The distillation procedure was as follows: 100 ml of each sample solution was spiked with *n*-butanol to give a 0.8 mg litre^{-1} level and saturated with 21.3 g of dried sodium sulphate in a 250 ml round bottomed flask. A 7.75 in Vigreux column was connected to the flask. A small distilling head with a thermometer and a small condenser attached was connected to the upper end of the Vigreux column. During the distillation process, the vapour temperature at the distilling head was controlled at 100 °C by regulating the voltage supply to the heating mantle. Distillation continued until the first 1.5 ml of distillate was collected. From that amount, 1 ml was injected with a 2 ml syringe into a 15 ml head-space serum bottle containing 1.2 g dried sodium sulphate. Distillations were done in duplicate for each sample. The standard solutions and distilled water were distilled in the same way.

Head-space bottles were heated in a 70 °C water bath for at least 70 min. The head-space gas was injected via an eight-port gas sampling valve into a gas chromatograph equipped with flame ionization detectors. The gas chromatograph was fitted with a 183 cm 2 mm i.d. glass column packed with 0.4 per cent Carbowax 1500 on 60/80 mesh Carbopack A (Supelco. Inco., Bellefonte, Pa.). After injection the column temperature was held at 60 °C for 2 min, then increased at 8 °C min^{-1} to 150 °C and held for 5 min. The carrier helium flow rate was 10 ml min^{-1}. An electronic integrator was interfaced to the flame ionization detector for acquisition of retention time and peak area data. The ratios of the peak areas of the resulting chromatograms to internal standard peak area were used for standard curves.

Heyndrickx and Peteghem[92] identified volatile components in waste waters by combined head-space analysis and gas chromatography. Drodz and Novak[93] have shown that low concentrations of volatile organics can be determined in water by the double sampling method of head-space gas chromatography using a closed loop strip/trap technique. Fielding *et al.*[94] have developed the following procedure of dynamic head-space analysis.

Carry out the extraction in the apparatus shown in Fig. 10. Place the sample (2 litres) in a precleaned quinol glass stoppered bottle in the extraction flask and add a solution (2 μl) of the internal standard (100 μg μl^{-}) in acetone. Use deuterated internal standards to provide quantitative information on some of the compounds identified, chlorobenzene-d_5, *p*-xylene-d_{10}, phenol-d_5, naphthalene-d_8, hexadecane-d_{34}, and phenanthrene-d_{10} added to each sample immediately before extraction are suitable. Place the flask in the water bath (thermostatted at 28 °C) and leave for 30 min. Then switch on the pump, and continue the purging (1.5 litres min^{-1}) for 3 h. Switch off the pump, remove the filter tube from the filter holder and connect to an eluate collection tube (Fig. 10 with PTFE tubing. Connect a short length (2 cm) of capillary tubing (5.5 mm o.d., 1.0 mm i.d.) to the other end of the filter tube to prevent evaporation of the solvent while extracting the filter. Elute the carbon filter with purified methylene chloride (2 × 15 μl, 1 × 20 μl) to give an extract suitable for gas chromatography—mass spectrometric analysis.

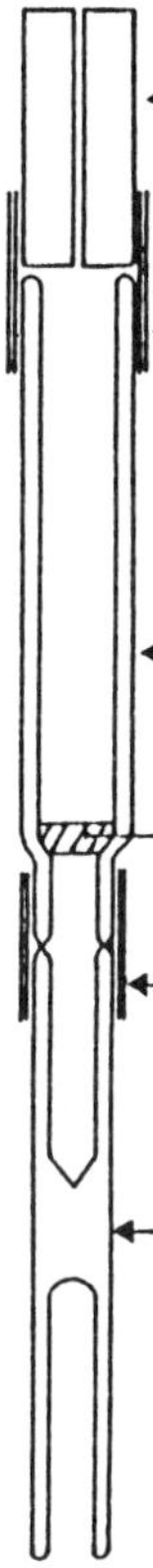

Fig. 10 Apparatus for elution of organic compounds from carbon adsorption filter after dynamic head-space extraction. From Fielding *et al.*[94] with permission.

A further refinement of head-space analysis is called closed loop stripping analysis. In this technique the sample at a controlled temperature is purged with inert stripping gas in a closed system whereby the gas is recycled through the sample continuously for a period of 0.5–2 h. In one version of the technique a sample is withdrawn from the cycling gas periodically and analysed directly for volatile organics by gas chromatography or gas chromatography—mass spectrometry. In another version of the apparatus the recyling gas passes continuously through a short column of a material such as Tenax GC which adsorbs organics or through a cryogenic trap. The column or trap is removed periodically and its contents analysed by gas chromatographic techniques.

Gomez *et al.*[135] optimized the conditions for the closed-loop stripping analysis (CLSA) method for the quantitative determination of volatile organics in water to reduce analysis time. Extraction efficiencies with respect to volatility, water temperature, and time were calculated for a series of 21 aliphatic and aromatic hydrocarbons and their chlorinated derivatives. Optimal recoveries were obtained by stripping at 45 °C for 0.5 h instead of the conventional 35 °C for 2.0 h. Operational procedures for overcoming background contamination were discussed. Janda *et al.*[136] used closed loop stripping analysis in conjunction with capillary column gas chromatography to identify unexpected substances appearing in the gas chromatograms of potable water following the addition of bromoalkanes as internal standards to the water sample. Due to the presence of free chlorine in the water sample bromoalkanes were converted to chloroalkanes which did not occur in unchlorinated water samples.

McNally and Grob[95] determined the solubility limits of organic priority pollutants by gas chromatographic head-space analysis. A simple and rapid method is described which accurately determines these limits. Head-space, or vapour equilibration analysis, may be used to determine the solubilities of the compound with even moderate volatility. As long as the compound shows some degree of volatility (vapour pressure), the method may be used to determine solubility. The method has been applied to chlorinated aromatics on the US Environmental Protection Agency's pollutants' lists. Otsen and Williams[96] obtained detection limits of 1 μg litre^{-1} and linear plots of peak area versus concentration in the range 0.25–16 μg litre^{-1} for 32 organic compounds examined by the head-space analysis technique. At concentrations between 4 and 16 μg litre^{-1} peak area precisions of 10 per cent RSD were obtained. Various applications of head-space analysis to the preconcentration of organics in natural waters are reviewed in Table 50.

6.2.2 Sea-water

May *et al.*[68] described a head-space gas chromatographic method for analysing hydrocarbons in sea-water which is sensitive at the sub-microgram per kilogram level. Dynamic head-space sampling for volatile hydrocarbon components, followed by coupled-column liquid chromatography for analysing the non-volatile components, requires minimal sample handling, thus reducing the risk of sample component loss and/or sample contamination. The volatile components are concentrated on a Tenax gas chromatographic precolumn and determined by gas chromatography or gas chromatography—mass spectrometry.

There have been many other applications of the technique.[123-132] The major advantage of the head-space method is simplicity in handling the materials. At most, only one chemical, the salt used in the salting-out procedure, needs to be added and in most cases the head-space gas can be injected directly into a gas chromatograph or carbon analyser. On the

Table 50 Preconcentration of organics from natural waters using head-space analysis.

Type of organic	Type of water sample	Sensitivity	Analytical finish	Ref.
Aliphatic hydrocarbons	Natural	1	GLC	97–100
Aromatic hydrocarbons	Natural	–	GLC	99,101,102
Halogenated paraffins	Natural	–	GLC	103–106
Tetrachloromethane, trichloroethylene	River	–	GLC	107
Chlorinated hydrocarbons	Industrial effluents	–	GLC–MS	102
Haloforms	Rain	0.001	GLC	108
Haloforms	Potable	–	GLC	109–120
Methyl bromide	Potable	–	GLC	121
Tetrahydrothiophen	Natural	–	GLC–MS	122

other hand, the concentration of organic materials present is limited by the volume of sea-water in the sample bottle. This is very much a batch process. Equilibration between the head-space gas and the solution can take a considerable time. This is not a problem when the salting-out material is added at sea, and the samples are then brought into the laboratory for analysis some time later. When the salting-out is done in the laboratory, equilibration can be hastened by recirculating the head-space gas through the solution. A system could be devised which would permit the accumulation of volatiles from a large volume of water into a relatively small head-space, perhaps by recirculating both water and head-space gas through a bubbling and collection chamber, but much of the simplicity and freedom from possible contamination would be lost in the process.

Volatile organic materials can also be removed from solution by distillation, either at normal or at elevated pressures. While the amounts to be collected in this fashion are small, if head-space samples are taken at elevated temperatures and pressures trace quantities of organics can be detected.[133] It should be emphasized, however, that whenever extreme conditions are employed to free an organic fraction, that fraction is defined by the conditions of the separation and cannot profitably be compared with fractions defined by different sampling conditions. The use of elevated temperatures and pressures may also alter the compounds separated, limiting the amount of information that can be extracted from the analyses.

Friant and Suffet[129] have investigated in detail the interactive effects of temperature, salt concentration, and pH on head-space analysis for isolating volatile trace organics in aqueous samples. Optimal conditions were derived from a statistical evaluation of the effect of parameter variation on the partition coefficient. These were a pH of 7.1, a sample temperature of 50 °C, and a salt concentration equivalent to 3.35 M sodium sulphate. Dowty *et al.*[126] passed the head-space purge gas through a column of Tenax GC (poly(*p*-2, 6-diphenylphenylene)oxide) adsorbent to trap the organics. The organics are then released from the Tenax GC and swept into a gas chromatograph for analysis in the parts per billion range. Bellar and Lichtenberg[128] also used the principle of adsorption of the organic components of the purge gas on a solid adsorbent material.

Chian *et al.*[132] point out that the Bellar and Lichtenberg[128] procedure of gas stripping followed by adsorption on to a suitable medium and subsequent thermal desorption on to a gas chromatograph—mass spectrometer is not very successful for trace determinations of volatile polar organic compounds such as the low molecular weight alcohols, ketones, and aldehydes. To achieve their required sensitivity of parts per billion Chian *et al.*[132] carried out a simple distillation of several hundred ml of sample to produce a few ml of distillate. This achieved a concentration factor of between 10 and 100. The head-space gas injection—gas chromatographic method was then applied to the concentrate obtained by distillation.

While the gases used in stripping are usually air, nitrogen or helium, electrolytically evolved hydrogen has been used as a collector for hydrocarbons.[134] In this technique, the gas is not passed through a column of adsorbent, but instead collects in the head space of the container. Since the volume of sea-water and of hydrogen are known, the hydrocarbon concentration in the head space can be used to calculate the partition coefficients and the concentrations of hydrocarbon in the sea-water.

References

1. Grob, K. (1973). *Journal of Chromatography*, *84*, 255.
2. Grob, K. and Grob, G. (1974). *Journal of Chromatography*, *90*, 303.
3. Grob, K. and Grob, G. (1975). *Journal of Chromatography*, *106*, 299.
4. Piet, S. J. (1975). Quarterly In *Report National Institute for Water Supply*, The Netherlands, No 2, 4.
5. Grob, K. and Zurcher, R. (1976). *Journal of Chromatography*, *117*, 285.
6. Obach, E. (1970). In *Organic Solvents*, eds J. A. Riddick and W. B. Burger, Wiley, Interscience, New York.
7. Kuo, P. P. K., Chian, E. S. K., DeWalle, F. B., and Kim, J. H. (1977). *Analytical Chemistry*, *49*, 1023.
8. Colenutt, B. A. and Thorburn, S. (1980). International *Journal of Environmental Analytical Chemistry*, *7*, 231.
9. Colenutt, B. A. and Thorburn, S. (1980). *International Journal of Environmental Studies*, *15*, 25.
10. Boren, H., Grimvall, A., Palmborg, J., Sauerhed, R., and Wigilius, B. (1985). *Journal of Chromatography*, *348*, 67.
11. Haberer, K. and Schredelslker, F. (1984). *Zeitschrift fur Wasser und Abwasser Forschung*, *17*, 206.
12. Westendorf, R. G. (1982). *International Laboratory*, *32*, September.
13. The Oil Companies International Study Group of Conservation of Clean Air and Water, Europe (1982). *Analysis of trace substances in aqueous effluents from Petroleum refineries*, CONCAWE report no 6/82.
14. In *Environmental Protection Agency Report EPA 601* (1980). Environmental Protection Agency, USA.
15. In *Environmental Protection Agency Report EPA 602* (1980). Environmental Protection Agency, USA.
16. Graydon, J. W., Grob, K., Zurcher, F., and Giger, W. (1984). *Journal of Chromatography*, *285*, 307.
17. Coleman, W. E., Munch, J. W., Slater, R. W., Melton, R. G., and Kopfler, F. C. (1983). *Environmental Science and Technology*, *17*, 571.
18. Blanchard, R. D. and Hardy, J. K. (1985). *Analytical Chemistry*, *57*, 2349.
19. Melcher, R. G. and Caldecourt, V. J. (1980). *Analytical Chemistry*, *52*, 875.

20. Bellar, T. A. (1974). *Journal of the American Water Works Association*, *566*, 739.
21. Dowty, B., Green, L., and Laster, J. T. (1976). *Journal of Chromatographic Science*, *14*, 187.
22. Dowty, B. and Land Laseter, J. L. (1975). *Analytical Letters*, *8*, 25.
23. Otson, R. and Chau, C. (1987). *International Journal of Environmental Analytical Chemistry*, *30*, 275.
24. Lopez-Avila, V., Heath, N., and Hu, A. (1987). *Journal of Chromatographic Science*, *25*, 356.
25. Lopez-Avila, V., Wood, R., Flanagan, M., and Scott, R. (1987). *Journal of Chromatographic Science*, *25*, 286.
26. Mosesman, N. H., Sidisky, L. M., and Corman, S. D. (1987). *Journal of Chromatographic Science*, *25*, 351.
27. Duffy, M., Driscoll, J. N., Pappas, S., and Sanford, W. (1988). *Journal of Chromatography*, *441*, 73.
28. Chichester-Constable, D. J., Barbean, M. E., Liu, S. E., Smith, S. R., and Stuart, J. D. (1987). *Analytical Letters*, *20*, 403.
29. Colenutt, B. A. and Thorburn, S. (1980). *International Journal of Environmental Studies*, *15*, 25.
30. Kaiser, R. (1971). *Journal of Chromatographic Science*, *9*, 227.
31. Bellar, T. A. and Lichtenberg, J. J. (1974). In *Environmental Monitoring Series EPA-670/4-74-009. The determination of volatile organic compounds at the $\mu g l^{-1}$ level in water by gas chromatography*. US Environmental Protection Agency.
32. Hrivnak, J. and Hassler, J. (1976). *Vodni Hospodarstvi*, *26*, 193.
33. Drodz, J., Vadakova, Z., and Novak, J. (1986). *Journal of Chromatography*, *354*, 47.
34. Epstein, P. S., Hauer, T., Wagner, M., Chase, S., and Giles, B. (1987). *Analytical Chemistry*, *59*, 1987.
35. Narang, R. S. and Bush, B. (1980). *Analytical Chemistry*, *52*, 2076.
36. Leggett, D. C. (1977). *Analytical Chemistry*, *49*, 880.
37. Thiela, D. R., Foreman, P. S., Davis, A., and Wyeth, P. (1987). *Environmental Science and Technology*, *21*, 145.
38. Mehran, M. F., Golkar, N., Mehran, M., and Cooper, W. J. (1988). *Journal of High Resolution Chromatography and Chromatography Communications*, *11*, 610.
39. Krasner, S. W. (1988). *Water Quality Bulletin*, *13*, 78.
40. Dravnieks, A., Krotoszynski, S. K., Whitfield, J., O'Donnell, A., and Burgwald, J. (1971). *Environmental Science and Technology*, *5*, 1220.
41. Bertsch, W., Anderson, E., and Hysler, G. (1975). *Journal of Chromatography*, *112*, 701.
42. The Oil Companies International Study Group for Conservation of Clean Air and Water, Europe, (1982). In *Analysis of Trace Substances in Aqueous Effluents from Petroleum Refineries*, Concawe Report, No 6/82.
43. Colenutt, B. A. and Thorburn, S. (1980). *International Journal of Environmental Analytical Chemistry*, *7*, 231.
44. Colenutt, B. A. and Thorburn, S. (1980). *International Journal of Environmental Analytical Chemistry*, *15*, 25.

45. Grob, K. and Zurcher, F. (1976). *Journal of Chromatography*, *117*, 285.
46. Grob, K. (1973). *Journal of Chromatography*, *84*, 255.
47. Grob, K. and Grob, G. (1974). *Journal of Chromatography*, *90*, 303.
48. Grob, K. and Grob, G. (1975). *Journal of Chromatography*, *106*, 249.
49. Westerdorf, R. G. (1982). *International Laboratory*, September, 32.
50. Waggott, A. and Reid, W. J. (1977). In, *Separation of Non-Volatile Organic Carbon in Sewage Effluent Using High Performance Liquid Chromatography*, Part I, Technical Report No 52, Water Research Centre, Stevenage, Herts.
51. Swinnerton, J. W. and Linnenbom, V. J. (1967). *Science*, *156*, 1119.
52. Bellar, T. A. and Lichtenberg, J. J. (1975). In, *The Determination of volatile organic compounds at the* $\mu g l^{-1}$ *level in water by gas chromatography*, US National Technical Information Service Report No. 237973/3GA, 30 pp.
53. Hammers, W. D. and Bosman, H. F. P. M. (1986). *Journal of Chromatography*, *360*, 425.
54. Bellar, T. A. and Lichtenberg, J. J. (1974). *Journal of American Water Works Association*, *66*, 739.
55. Quimby, B. D. and Delaney, M. F. (1979). *Analytical Chemistry*, *51*, 875.
56. Quimby, B. D., Uden, P. C., and Barnes, P. C. (1978). *Analytical Chemistry*, *50*, 2112.
57. Chibar, K. and Haraguchi, H. (1983). *Analytical Chemistry*, *55*, 1504.
58. Kirschen, N. A. (1981). *Varians Instrument Applications*, *15*, 2.
59. *Federal Register (US)* 69464, *44*, No. 223, 3 December (1979).
60. Pfaender, F. E., Jones, R. B., Stevens, A. M., Moore, L., and Hass, J. R. (1978). *Environmental Science and Technology*, *12*, 438.
61. Dressman, R. C., Stevens, A. A., Fair, J., and Smith, B. (1979). *Journal of American Water Works Association*, *71*, 392.
62. Brass, H. J. (1980). *American Laboratory*, *12*, 23.
63. Kirshen, N. (1984). *American Laboratory*, *16*, 60.
64. Bellar, T. A., Lichtenberg, J. J., and Eichelberger, J. W. (1976). *Environmental Science and Technology*, *10*, 926.
65. Perras, J. C. (1973). In, *Portable Gas Chromatographic Technique to Measure Dissolved Hydrocarbons in Seawater*, USNTIS, AD Report No 786583/5GA.
66. Swinnerton, J. W. and Linnenbom, V. J. (1967). *Science*, 156, 1119.
67. Bellar, T. A. and Lichtenberg, J. J. (1975). In, *The Determination of Volatile Organic Compounds at the* μ*gl Level in Water by Gas Chromatography*, US National Technical Information Service Report No 237973/3GA.
68. May, W. E., Chester, S. N., and Cram, S. P. (1975). *Journal of Chromatographic Science*, *13*, 555.
69. MacKinnon, M. D. (1977). In, *The Analysis of the Total Organic Carbon in Seawater (a) Development of Methods for the Quantification of TOC, (b) Measurement and Examination of the Volatile Traction of the TOC.* PhD Thesis, Dalhouse University, pp. 1–183.
70. Rayer, J. and Hennequin, C. (1982). *Techniques et Sciences Minucipales*, *77*, 25.

71. Bodings, H. T., De Jong, C., and Dooper, R. P. M. (1985). *Journal of High Resolution Chromatography and Chromatography Communications*, *8*, 755.
72. Swinnerton, J. W. and Linnenbom, V. J. (1976). *Journal of Gas Chromatography*, *5*, 570.
73. Novak, J., Zluticky, J., Kubela, V., and Mostecky, J. (1973). *Journal of Chromatography*, *76*, 45.
74. Smith, R. (1987). *Journal of High Resolution Chromatography and Chromatography Communications*, *10*, 60.
75. Cochrane, J. W. (1987). *Journal of High Resolution Chromatography*, *10*, 573.
76. Gryder–Boutet, D. E. and Kennish, J. M. (1988). *Journal of American Water Works Association*, *80*, 52.
77. Cailleux, A., Turcant, A., Allain, P., Toussaint, D., Gaste, J., and Roux, A. (1987). *Journal of Chromatography*, *391*, 280.
78. Henatsch, J. J. and Juttner, F. (1988). *Journal of Chromatography*, *445*, 97.
79. Novak, J., Zluticky, J., Fubelka, V., and Mostocky, J. (1973). *Journal of Chromatography*, *76*, 45.
80. Brooks, J. M. and Sackett, W. M. (1973). *Journal of Geophysical Research*, *78*, 5248.
81. Atkinson, L. P. and Richard, F. A. (1967). *Deep Sea Research*, *14*, 673.
82. Cerwen, J. P. (1970). Volatile Organic Materials in Seawater. In, *Organic Matter in Natural Waters*, ed W. Hood, Institute of Marine Science, University of Alaska Publication No. 1, pp. 169–180.
83. Kaiser, K. L. E. and Oliver, B. G. (1976). *Analytical Chemistry*, *48*, 2207.
84. Wyllie, P. L. (1988). *Journal of American Water Works Association*, *80*, 65.
85. Munz, C. and Roberts, P. V. (1987). *Journal of the American Water Works Association*, *79*, 62.
86. Friant, S. L. and Suffet, I. (1979). *Analytical Chemistry*, *51*, 2167.
87. Malten, L. and Vreden, N. (1978). *Forum Stadte Hygiene*, *29*, 37.
88. Cowen, W. F., Cooper, W. J., and Highfill, J. W. (1975). *Analytical Chemistry*, *47*, 2483.
89. Gjavotchanoff, S., Luessem, H., and Schlimme, K., (1971). *Gas-u-Wasser Fach*, *112*, 448.
90. Chian, E. S. K., Kuo, P. P. K., Cooper, W. J., Cowen, W. F., and Fuentes, R. C. (1977). *Environmental Science and Technology*, *11*, 282.
91. Cowen, W. F., Cooper, W. J., and Highfill, W. J. (1975). *Analytical Chemistry*, *47*, 2483.
92. Heyndrickx, A. and Peteghem, C. (1975). *European Journal of Toxicology*, *8*, 275.
93. Drodz, J. and Novak, J. (1982). *International Journal of Environmental Analytical Chemistry*, *11*, 241.
94. Fielding, M., Gibson, T. M., James, H. A., McLoughlin, K., and Steel, C. P. (1981). In *Technical report TR 159G. Organic Micropollutants in drinking water*. Water Research Centre, Medmenham, UK, February.

95. McNally, M. F. and Grob, R. L., (1983). *Journal of Chromatography*, *260*, 23.
96. Otsen, R. and Williams, D. T. (1982). *Analytical Chemistry*, *54*, 942.
97. Khazal, W. J., Vejrostra, J., and Novak, J. (1978). *Journal of Chromatography*, *157*, 125.
98. Drodz, J. and Novak, J. (1978). *Journal of Chromatography*, *152*, 55.
99. Drodz, J., Novak, J., and Rijks, J. (1978). *Journal of Chromatography*, *158*, 471.
100. McAuclliffe, C. J. (1971). *Chemical Technology*, *1*, 46.
101. Kaiser, L. E. and Oliver, B. G. (1976). *Analytical Chemistry*, *48*, 2207.
102. Schulz, J. (1987). *Vom Wasser*, *69*, 49.
103. Kaiser, K. L. and Oliver, B. G. (1976). *Analytical Chemistry*, *48*, 2207.
104. Hellman, Z. (1985). *Zeitschrift fur Wasser und Abwasser Furschung*, *18*, 92.
105. Mehran, M., Cooper, M., and Jennings, W. (1986). *Journal of Chromatographic Science*, *24*, 142.
106. Kirshen, N. (1984). *American Laboratory*, *16*, 60.
107. Hellmann, H. (1988). *Zeitschrift fur Wasser und Abwasser Forschung*, *21*, 67.
108. Comba, M. E. and Kaiser, K. L. (1983). *International Journal of Environmental Analytical Chemistry*, *16*, 17.
109. Kaiser, K. L. E. and Oliver, B. G. (1976). *Analytical Chemistry*, *48*, 2107.
110. Deitz, E. A. and Singley, K. F. (1979). *Analytical Chemistry*, *51*, 1809.
111. Otson, R., William, D. T., and Bothwell, P. D. (1979). *Environmental Science and Technology*, *13*, 936.
112. Varma, M. M., Siddigue, M. R., Doty, K. T., and Machis, H. (1979). *Journal of American Water Works Association*, *71*, 389.
113. Symons, J. M., Bell, T. A., Carswell, J. K., DeMarco, J., Krupp, L. L., Robeck, G. G., Seegar, D. R., Slocum, D. J., Smith, B. L., and Stevens, A. A. (1975). *Journal of American Water Works Association*, *67*, 643.
114. Keith, L. H. (1976). In *Identification and Analysis of organic pollutants in water*. Ann Arbor Science Publishers Ltd, Ann Arbor, Michigan.
115. In *National Survey for halomethanes in drinking water* (1977). Health and Welfare, Canada, 77,-EHD-9.
116. Bush, B., Narang, R. S., and Syrotynski, S. (1977). *Bulletin Environmental Contamination and Toxicology*, *13*, 436.
117. Morris, R. J. and Johnson, L. G. (1976). *Journal of American Water Works Association*, *68*, 492.
118. Castello, G., Gerbino, T. C., and Kanitz, S. (1986). *Journal of Chromatography*, *351*, 165.
119. Croll, B. T., Summer, M. E., and Leathard, D. A. (1986). *Analyst (London)*, *111*, 73.
120. Castello, G., Gerbino, T. C., and Kanitz, S. (1982). *Journal of Chromatography*, *247*, 263.
121. Cirilli, L. and Borgioli, A. (1986). *Water Research*, *20*, 273.
122. Gorlucci, G., Airoldi, L., and Farelli, R. (1984). *Journal of Chromatography*, *287*, 425.

123. Gjavotchanoff, S., Luessen, H., and Schlimme, E. (1971). *Gas and Wasser*, *112*, 448.
124. Chesler, S. N., Gump, B. H., Hertz, H. P. (1978). *Analytical Chemistry*, *50*, 805.
125. Heyndrickx, A. and Peteghem, C. (1975). *European Journal of Toxicology*, *8*, 275.
126. Dowty, B., Green, L., and Laseter, J. L. (1976). *Journal of Chromatographic Science*, *14*, 187.
127. Waggott, A. (1978). In, *Proceedings of the Analytical Division of Chemical Society*, London, *15*, 232.
128. Bellar, T. A. and Lichtenberg, J. J. (1974). *Journal of American Water Works Association*, *566*, 739.
129. Friant, S. L. and Suffet, I. H. (1979). *Analytical Chemistry*, *51*, 2161.
130. Malter, L. and Vreden, V. (1978). *Forum Stadte-Hygiene*, *29*, 37.
131. Cowen, W. F., Cooper, W. J., and Hisfill, J. W. (1975). *Analytical Chemistry*, *47*, 2483.
132. Chian, E. S. K., Kuo, P. P. K., Cooper, W. J., Cowen, W. F., and Fuentes, R. C. (1977). *Environmental Science and Technology*, *11*, 282.
133. Luessem, H. (1972). Detection of trace materials by concentration by pressure distillation. *Haus Tech., Essen, Vortragsveroeff.*, *1972*, 69.
134. Wasik, S. P. (1974). *Journal of Chromatographic Science*, *12*, 845.
135. Gomez, B., Chan, J. I., and Alaiges, J. (1987). *International Journal of Environmental Analytical Chemistry*, *30*, 183.
136. Janda, V., Marha, K., and Mitera, J. (1988). *Journal of High Resolution Chromatography and Chromatography Communications*, *11*, 541.

7

MISCELLANEOUS PRECONCENTRATION TECHNIQUES

There are a number of additional techniques which have been adapted for the preconcentration, principally, of cations and organics, as a means of increasing the sensitivity of analysis for these substances. Many of these techniques are of value in particular applications and are discussed here under various broad headings. Techniques that have been used for cations includes: electrolytic (Section 7.1.1); gas evolution (Section 7.1.2); flotation (Section 7.1.3, also used for anious); evaporation/distillation (Section 7.1.4); precipitation/crystallization (Section 7.1.5); and dialysis/osmosis (Section 7.1.6). In the case of organics the following techniques have been examined: electrolytic (Section 7.3.1); flotation (Section 7.3.2); evaporation (Section 7.3.3); freezing (Section 7.3.4); and osmosis (Section 7.3.5).

7.1 Metal cations

7.1.1 Electrolytic methods

Non-saline waters Methods have been devised for the following elements in water samples: lead;[1] chromium, cobalt, copper, mercury, nickel, and zinc,[2] silver, bismuth, cadmium, cobalt, chromium, mercury, indium, nickel, lead, selenium, tellurium, thallium, and zinc;[3] mercury;[4] and cadmium.[5] In one procedure, lead was preconcentrated by electrolysis on a silanized graphite rod. The atomization was performed by atomic heating of the rod in a hot helium flow. The lead was determined by dispersionless X-ray fluorescence spectromentry in amounts down to 15 pg.

Frick and Tallman[6] have described a flow cell for the determination of mercury in water by electrodeposition followed by atomic absorption spectrometry. They use commercially available non-coated graphite furnace tube (0,222 ± 0.0051 in i.d.) as the working electrode in a thin layer, flow-through configuration. The thin-layer design enhances deposition efficiency and hence sensitivity compared to deposition carried out with an open tubular electrode. Electrodepositions were performed with a PAR 174A polarographic analyser (EG & G Princeton Applied Research, Princeton, NJ) using a tubular graphite working electrode flow cell described below, a counter-electrode of 52 mesh platimum gauze 52 mm square contained in an isolation bridge, and a silver/silver chloride 3.5 M potassium chloride reference electrode. Solution was pumped through the flow cell using a Masterflex variable speed peristaltic pump equipped with a 7016 head (Cole-Parmer Instrument Co., Chicago, IL).

After mercury deposition on the graphite tube it was placed in a Perkin Elmer Model 603 spectrometer equipped with a HGA 2000 graphite tube furnace. The atomizer temperature program consisted of a 60 s dry time at 105 °C (with an argon gas flow through the furnace of 250 ml min^{-1}) and an 8 s atomization at 1600 °C (argon gas flow interrupted). Sharp, narrow atomization profiles were obtained. A standard curve for inorganic mercury was linear through 6 mg $litre^{-1}$ and an analytically useful curve was attained through 10 μg $litre^{-1}$. The relative standard deviation ranged from 12 per cent at the 0.2 μg $litre^{-1}$ level to 1.9 per cent at the 8 μg $litre^{-1}$ level, averaging 4.4 per cent over the entire range. Repeated analysis of a 4 μg $litre^{-1}$ test solution using five different graphite tubes chosen at random indicated a variation among tubes characterized by an RSD of 6 per cent. A detection limit ($\alpha = 2$) of 0.08 μg $litre^{-1}$ is estimated from the sensitivity (0.071 absorbance units per μg $litre^{-1}$) and the noise (RSD) associated with measuring the absorbance signal of a blank. The accuracy of the electrodeposition atomic absorption method was tested by analysing an Environmental Protection Agency reference sample (accepted mercury concentration of 7.6 μg $litre^{-1}$). The two methods yielded mean results of 7.3 and 7.8 μg $litre^{-1}$, respectively. A river water sample, after 0.45 μm filtration was also analysed by the electrodeposition atomic absorption spectrometric system and by the cold vapour method, with each technique giving the same value of 2.8 μg $litre^{-1}$.

Muhlbaier *et al.*[7] determined cadmium in Lake Michigan water by electrodeposition on an amalgamated gold wire followed by electrothermal atomic absorption spectrometry or by mass spectrometric isotope dilution analysis used as a reference method. This process achieved a 5000-fold enrichment in cadmium content. The sample container—electolysis cell used by these workers is shown in Fig. 11. Prior to use, the entire apparatus was cleaned with a phosphate free detergent, rinsed, leached with 10 per cent v/v aqueous Ultrex ultrapure nitric acid and then rinsed and filled with Nanopure water and placed in double plastic bag. After connection to the outlet of the sampling device the container was flushed with several times its volume of sample before a subsample was entrained by closing the stopcocks. As shortly thereafter as feasible, the sample was treated with 1 ml of 10 per cent v/v nitric acid.

The cell's anode consisted of a 30 ml diameter high purity platinum wire 10 cm long. The cathode was 30 ml gold wire of the same length, amalgamated by brief immersion in triple distilled mercury. Because the

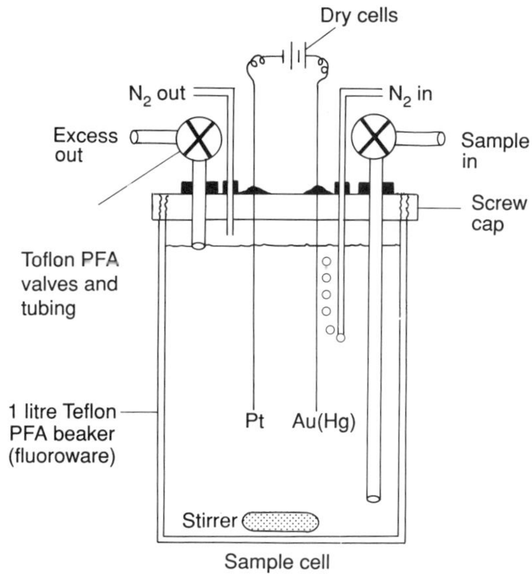

Fig. 11 Schematic representation of combination sample container and electrodeposition cell. From Muhlbaier *et al.*[7] with permission.

mercury is incorporated into the concentrate during stripping, its purity must be ascertained in advance to avoid introducing spurious cadmium. Three 1.5 V dry cells connected in series provided the EMF for deposition. This arrangement resulted in a current of 2–3 mA.

After electrodeposition, the cathode and anode were separately stripped with 200 μl of 10 per cent v/v Ultrex nitric acid contained in a micro test tube made by sealing one end of a length of 1 mm capillary tubing. The strippings were analysed by atomic absorption spectrometry or by the mass spectrometric reference method.

Muhlbaier *et al.*[7] conclude that recoveries obtained by atomic absorption spectrometry are appreciably lower than those obtained by the mass spectrometric reference method (Table 51). The latter method was suitable for determining nanogram levels of cadmium in fresh-water samples. Once the proper precautions are taken, the procedure yields a concentrate that is free of significant interferences and which gives steady ion currents from samples and blanks containing 25–100 ng total cadmium. In practice the lower working limit is determined by the blank value, rather than inherent sensitivity and it is the variability introduced by contamination that appears to control the overall precision. Clean room conditions would definitely be required if metal concentrations below 10 μg litre^{-1} were being determined.

Szy and Toth[8] preconcentrated uranium-234 and uranium-238 on

Table 51 Cadmium contents of Lake Michigan samples by two techniques.

	Cadmium content ± SD (ng litre^{-1})	
	Sample A	Sample B
Electrodeposition–atomic absorption spectrometry	24 ± 8	25 ± 2
Electrodeposition–mass spectrometric isotope dilution analysis	38 ± 8	31 ± 8
Per cent recovery obtained in atomic absoprtion method relative to mass spectrometric referee method	63.2	80.6

From Muhlbaier *et al.*[7] with permission.

a stainless steel disc prior to carring out scintillation measurement. Coprecipitation with ferric hydroxide has been used to preconcentrate uranium-234 and uranium prior to α-scintillation spectrometry. Trettenbach and Henmann[9] used cathodic deposition for the initial preconcentration of lead, cadmium, and thallium from potable water prior to their determination at the μg litre^{-1} and ng litre^{-1} ranges by isotope dilution mass spectrometry.

Wang and Mahmoud[10] described preconcentration procedure for ultra trace quantities of iron(lI) based on the effective interfacial accumulation of iron(III)/Solochrome Violet RS (SVRS) chelate on the hanging mercury drop electrode. After degassing, a preconcentration potential of − 0.40 V was applied while the solution was stirred. After concentration, the stirring was stopped and the voltammogram recorded to a final potential of −1.10 V. The height of the chelate peak which occurred at a potential of 0.28 V more negative than the peak of the free dye was proportional to the concentration of the iron. The limit of detection at a solution pH of 5.1, after 1 min preconcentration was 0.04 μg litre^{-1} and the relative standard deviation at the 0.1 μM level was 4.7 per cent. The effects of interferences due to coexisting metal ions or organic surfactants were evaluated. Iron (III) could be determined in the presence of iron(II).

Prabhu *et al.*[11] studied the preconcentration of copper(I) at a chemically modified carbon paste electrode containing 2,9-dimethyl-1, 10-phenanthroline. Deposition was rapid with 80–90 per cent of the final response generated within 1 min of exposure to copper(I) solution. Quantification of copper(I) from the the area of the anodic curve for a 1 s deposition time gave a limit of detection of 0.3 μM and a linear response up to 10 μM. Relative standard deviation was 5 per cent. A small preconcentration signal was observed for silver(I), but there were no direct responses for copper(II), zinc(II), cobalt (II), lead(II), nickel(II), or iron (III), although concentrations of these suppressed the copper(I) response.

Sea-water For most tasks in the trace chemistry of natural waters, voltammetric determination requires preconcentration, because in a group of simultaneously determined ecotoxic heavy metals one usually has levels below 0.1 μg litre^{-1}. Electrochemical preconcentration can be attained in the following two ways, depending whether differential pulse stripping voltammetry (differential pulse anodic scanning voltammentry) or adsorption differential pulse voltammentry has been applied.

Heavy metals capable of forming amalgams, i.e. Cu, Pb, Cd, Zn, etc., are plated at a stationary mercury electrode consisting of a hanging mercury drop electrode with adjustment of a rather negative potential of -1.2 V versus the Ag/AgCl reference electrode for several minutes. To speed up mass transfer, the solution is stirred with a magnetic bar at 900 r.p.m. Their preconcentration is achieved by the accumulation of the heavy metals in the mercury drop. Subsequently the stirring is terminated, and after a quiescent period of 30 s the potential is scanned into the anodic direction in the differential pulse mode. At the respective redox potential the plated heavy metal is reoxidized and the corresponding current is recorded (Fig. 12). The voltammetric peak heights obtained are proportional to the bulk concentrations of the respective trace metals in the analyte. The hanging drop mercury electrode can usually be applied down to trace levels of 0.1–0.05 μg litre^{-1}.

At lower ultra-trace levels the less voluminous mercury film electrode has to be used. It consists of a mercury film of only several hundred Å thickness on a glassy carbon electrode as support. The fabrication of this glassy carbon electrode is critical for obtaining an optimal mercury film electrode suitable to perform determinations down to 1 ng litre^{-1} or below.

Certain trace substances such as Se^{IV} can be determined in a similar manner by differential cathodic stripping voltammetry (DPCSV). For selenium a rather positive preconcentration potential of -0.2 V is adjusted. Se^{IV} is reduced to Se^{2-} and Hg from the electrode is oxidized to Hg^{2+} at this potential. It forms, with Se^{2-} on the electrode, a layer of insoluble HgSe and in this manner the preconcentration is achieved. Subsequently the potential is altered in the cathodic direction in the differential pulse mode. The Hg^{II} resulting peak produced by the Hg^{II} reduction is proportional to the bulk concentration of Se^{IV} in the analyte. Cathodic stripping voltammetry has been discussed by Van der Berg.[12] This procedure has recently made rapid progress and could now be used to determine lead, cadmium, copper zinc, uranium, vanadium,

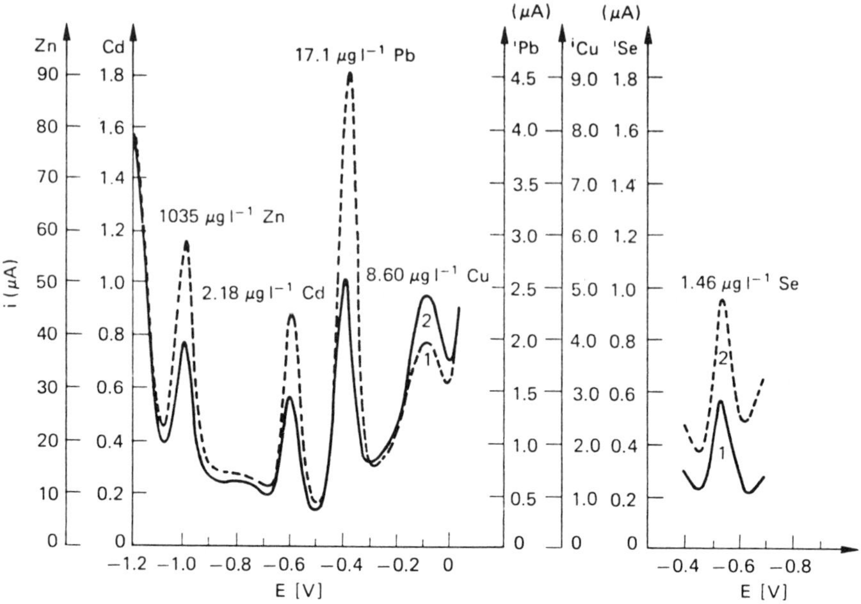

Fig. 12 Voltammogram of the simultaneous determination of Cu, Pb, Cd, and Zn with DPASV at the HMDE and subsequent determination of Se^{IV} by DPCSV in the same run in rain water at an adjusted pH of 2. Preconcentration time for DPASV is 3 min at −1.2V, for DPCSV 5 min at −0.2V. 1 = Original analyte; 2 = after first standard addition. Total analysis time with two standard additions, 30–40 min.

molybdenum, nickel, and cobalt in water, with great sensitivity and specificity, allowing study of metal speciation directly in the unaltered sample. The technique used preconcentration of the metal at a higher oxidation state by adsorption of certain surface-active complexes, after which its concentration was determined by reduction. The reaction mechanisms, effect of variation of the adsorption potential, maximal adsorption capacity of the hanging mercury drop electrode, end possible interferences are discussed.

Examples of extraction of trace metals from sea-water as chelates with subsequent determination by electrothermal atomic absorption spectrometric procedures have been described[13-16] but these, and similar methods, are seldom effective and satisfactory when the matrix is very complex and the analyte concentration very low. In contrast, the coupling of electrochemical and spectroscopic techniques, i.e. electrodeposition of a metal followed by detection by atomic absorption spectrometry, has received limited attention. Wire filaments, graphite rods, pyrolytic graphite tubes, and hanging drop mercury electrodes have been tested[17-28] for electrochemical preconcentration of the analyte to be determined by atomic absorption spectrometry. However, these *ex-situ* preconcentration methods are often characterized by unavoidable irreproducibility, contaminations arising from handling of the support, and detection limits unsuitable for lead detection at sub-ppb levels.

These drawbacks could be certainly avoided by performing *in-situ* deposition. The sole attempt in this direct ion was made by Torsi[29] who set up an apparatus which permitted both *in-situ* electrochemical preconcentration of the analyte from a flowing solution and almost complete suppression of matrix effects because the matrix could be removed by suitable washing. The feasibility of this approach was successfully tested with respect to lead determinations in different salt solutions (mainly ammonium acetate) and some preliminary results were reported for real samples, such as sea-water and urine.[29]

Torsi *et al.*[29,30] have carried out a systematic investigation to establish the potentialities of such an apparatus. The apparatus is basically an electrothermal device in which the furnace (or the rod) is replaced by a small crucible made of glassy carbon. Figure 13(a) represents an overall view of the apparatus. Figure 13(b) shows a block diagram of the electrolysis circuit; the crucible (6) acts as a cathode while the anode is a platinum foil dipped into either the sample solution reservoir (1) or the washing solution reservoir (2). The pre-electrolysis was performed at constant current by a 500 V DC variable power supply (5). Under these conditions, the cathode potential is not controlled so that other metals can be co-deposited with lead. There is no great need to control the deposition potential, because the spectral selectivity is sufficiently good to prevent interferences by other metals during the atomic absorption step.

A typical measurement was performed as follows. The feeder was lowered into the crucible and the sample solution (sea-water) was allowed to flow under an inert atmosphere with the suction on. A constant current was applied for a predetermined time. When the pre-electrolysis was over, the flow was changed from the sample to the ammonium acetate washing solution, while the deposited metals were maintained under cathodic protection. Ammonium acetate was selected for its low decomposition temperature, and a 0.2 mg litre^{-1} concentration was used to ensure sufficient conductivity. At this point the feeder tip was raised to the highest position and the usual steps for an electrothermal atomic absorption spectrometry measurement were followed: drying for 30 s at 90 °C, ashing for 30 s at 700 °C, and atomization for 8 s at 1700 °C with measurement at 283.3 nm. A typical atomization signal obtained in this way is shown in Fig. 14(a). As can be seen, the baseline increases smoothly with time as a consequence of an upward lift of the crucible caused by thermal expansion of the material.

A calibration curve for lead in sea-water obtained by the standard

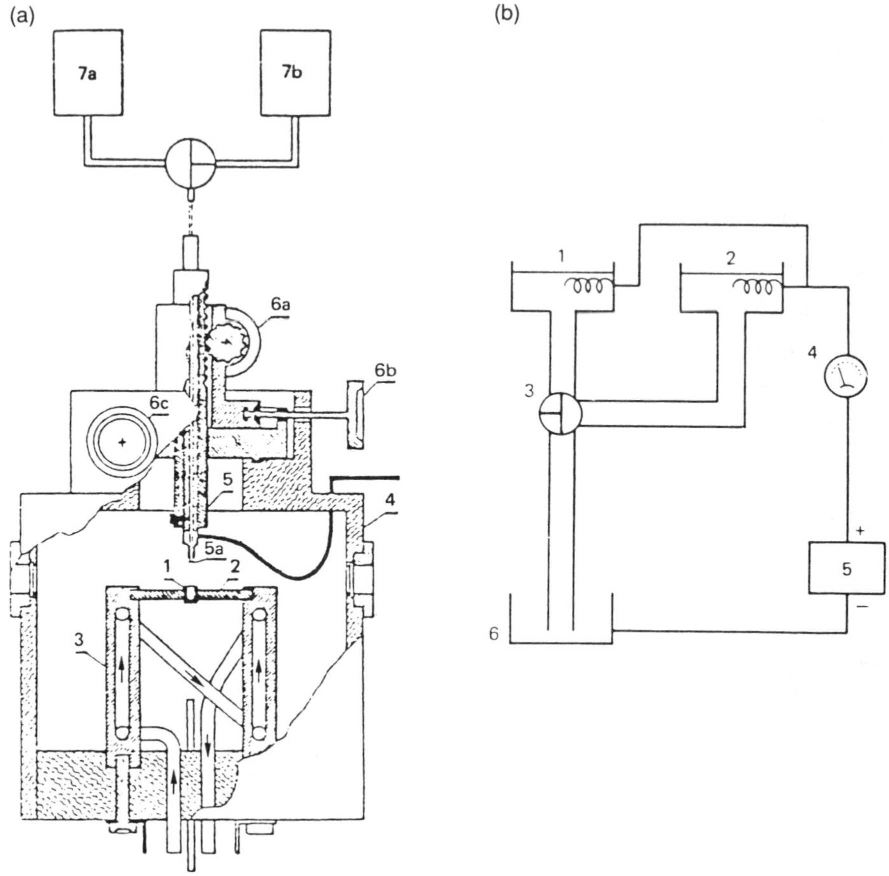

Fig. 13 (a) Overall view of the apparatus. 1 = Vitreous carbon crucible; 2 = graphite rod; 3 = water-cooled, steel column electrical leads; 4 = plexiglas cover; 5 = feeder; 5a = feeder tip; 6(a–c) = slide knobs; 7(a,b) = washing and sample solution reservoirs. The glassy carbon crucible (1) was 8 mm high, 5 mm o.d., 3 mm i.d., 6 mm deep (Le Carbon Lorraine, Paris), graphite rods (2) keep the crucible firmly in position. Water-cooled stainless steel columns (3) press the graphite rods against the crucible by two screws hidden inside and act also as electrical leads. The plexiglas box (4) allowed the use of the controlled inert atmosphere necessary to avoid drastic reduction of the absorption signal caused by oxygen. The solution feeder (5) can be moved up and down by means of a knob (6a) into a metal block attached to the upper part of the plexiglas box. A three-way stop-cock at the cylinder top connects, by Teflon tubing (1.5 mm o.d., 0.8 mm i.d.,) the feeder tip (5a) to the washing and sample solution reservoirs (see below). Other knobs (6b and c) enable the feeder to be moved in the horizontal plane. The three knobs permit a micrometric spatial adjustment of the feeder tip. (b) Electrolysis circuit layout. 1, 2 = Sample and washing solution compartment; 3 = three-way stopcock; 4 = ammeter; 5 = 500 V DC variable power supply; 6 = crucible. From Torsi[29] with permission.

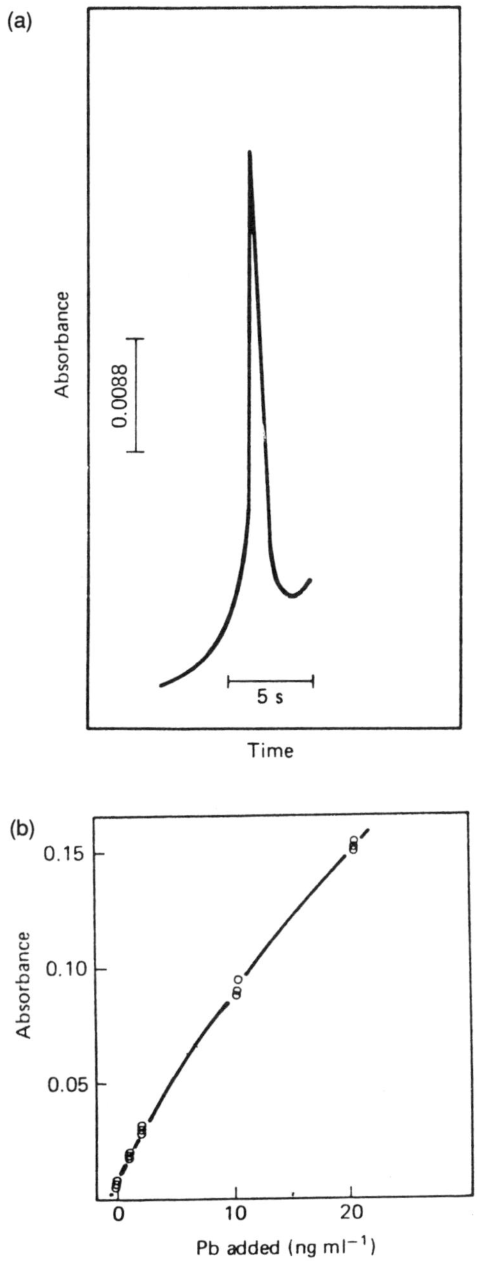

Fig. 14 (a) Typical recorder trace of sea-water containing 2.8 ng Pb^{2+} per ml after a 2 min electrolysis time. (b) Calibration curve for lead in sea-water, pH 1.9. From Torsi[29] with permission.

addition method is shown in Fig. 14(b), a deviation from linearity is observed at higher lead concentrations. The estimated value for the original sample was found to be 0.51 μg litre^{-1} with confidence limits at the 95 per cent confidence level of ±0.036 μg litre^{-1}, compared with a value of 0.65 ± 0.08 μg litre^{-1} obtained by anodic scanning.

Cowen, *et al.*[31] showed that polonium can be electrodeposited on to carbon rods directly from acidified sea-water, stripped from the rods, and autoplated onto silver counting discs with an overall recovery of tracer of 85 ± 4 per cent an electrodeposition time of 16 h.[32] These workers compared two procedures for concentrating polonium-210 from sea-water:

1. Coprecipitation upon partial precipitation of the natural calcium and magnesium with sodium hydroxide.
2. Electrodeposition of polonium directly from acidified sea-water on to carbon rods.

Table 52 Comparison of sodium hydroxide precipitation and carbon rod plating methods for concentrating ^{210}Po from aliquots of acidified sea-water removed from 50-litre parent samples.

Method	Date sampled	^{210}Po activity (pCi litre^{-1})[a]	Recovery (per cent)
Sample 1 (0.5 M HCl)[b]			
NaOH	12/5/75	0.115 ± 0.009	71
NaOH	12/5/75	0.113 ± 0.009	66
Carbon rod	12/5/75	0.116 ± 0.007	81[c]
Carbon rod	12/5/75	0.110 ± 0.005	89[c]
NaOH	17/5/75	0.104 ± 0.008	74
NaOH	17/5/75	0.097 ± 0.008	80
Sample 2 (0.5 M HCl)[d]			
NaOH	20/5/75	0.031 ± 0.002	86
NaOH	20/5/75	0.025 ± 0.002	74
Carbon rod	20/5/75	0.034 ± 0.002	63[e]
Carbon rod	20/5/75	0.035 ± 65[e]	
Carbon rod	27/5/75	0.034 ± 0.003	38[f]
Carbon rod	27/5/75	0.028 ± 0.002	41[f]
NaOH	27/5/75	0.040 ± 0.003	73
NaOH	27/5/75	0.034 ± 0.002	90

[a] ± 1 standard counting error.
[b] Collected on 25/4/75 at Scripps Pier.
[c] Electroplating time, 48 h.
[d] Collected on 20/5/75 at Scripps Pier.
[e] Electroplating time, 24 h.
[f] Electroplating time, 16 h.
From Cowen *et al.*[31] with permission.

Polonium thus concentrated was autoplated on to silver counting discs held in spinning Teflon holders. A comparison of results obtained by the two methods is shown in Table 52. Recoveries of polonium-208 tracer in the precipitation method were 77 ± 7 per cent ($n + 8$) compared with 40 ± 2 per cent ($n = 2$) for the electrodeposition method with 16 h plating time, 64 ± 1 per cent ($n = 2$) in 24 h, and 85 ± 4 per cent ($n = 2$) in 48 h. Even though the electrode-position method requires less attention, it requires long plating times for high recoveries. Thus the recovery of polonium-210 by direct plating appears to be rate limited by diffusion of polonium to the cathodes since the applied potential difference is far in excess of that required to reduce Po^{IV} to polonium.

Recoveries are based on the added polonium-208 tracer, presumably as $PoCl_6^{2-}$. Equilibrium was assumed to have occurred during the 2 h mixing of spike and sample which was always 0.5 M in acid at this stage. Table 53 shows that recovery of polonium-208 from sea-water was insensitive to the amount of 1 M sodium hydroxide used in precipitation. Sodium hydroxide 6 ml 1 M was chosen because it gave an easily manipulated volume of precipitate.

The need to acidify sea-water to be used for polonium analysis was investigated by periodically sub-sampling two parent samples over a week. Two 50 litre polyethlene carboys were filled with sea-water on the same date at the same location. One was acidified with 12 M HCl to 0.5 M. The 50-litre carboys were shaken thoroughly before water was removed. The polonium-210 concentration in the unacidified sea-water showed a dramatic decrease to less than half in 7 days, the time sometimes lapsing between field sampling, and analysis in the laboratory. Since the unacidified sea-water sample was acidified to 0.5 M on addition of the polonium-208 tracer, polonium-210 was either lost to the walls of the carboy or converted to a suspended species not leachable in the 0.5 M acid.

Another approach to the effect of acidification was to subsample a large bottle of sea-water, first untreated with acid and then progressively

Table 53 Effect of added sodium hydroxide on the recovery of ^{208}Po tracer from seawater[a].

Volume of 1N NaOH (ml litre^{-1})	Recovery (per cent)
3	86
6	83
10	86

[a] After neutralization of acid used to stabilize polonium in solution (0.5 M HCl).
From Cowen *et al.*[31] with permission.

Table 54 Effect of acidity and time on ^{210}Po concentrations in aliquots removed from single 100 litre parent sample[a].

Method	Date sampled	pCi litre^{-1} [b]	Recovery (per cent)
Raw seawater			
NaOH	29/5/75	0.097 ± 0.006	62
Carbon rod[c]	29/5/75	0.044 ± 0.003	74
NaOH	2/6/75	0.028 ± 0.002	86
Carbon rod[c]	2/6/75	0.047 ± 0.003	53
0.1 M HCl			
Carbon rod[c]	2/6/75	0.053 ± 0.003	64
Carbon rod[c]	4/6/75	0.050 ± 0.004	50
0.5 M HCl			
NaOH	10/6/75	0.052 ± 0.005	72
Carbon rod[c]	10/6/75	0.060 ± 0.006	55
NaOH	23/6/75	0.050 ± 0.002	82
Carbon rod[c]	23/6/75	0.058 ± 0.002	60

[a] Collected on 28/5/75 at Scripps Pier.
[b] ±1 standard counting error.
[c] Electroplating time, approximately 24 h.
From Cowen *et al.*[31] with permission.

made more acidic, allowing 2 or more days between treatments. The results of such a test are summarized in Table 54. The need for acidification on collection is again demonstrated and the large variations encountered in analysing raw sea-water by the sodium hydroxide precipitation method are evident.

Breyer and Gilbert[33] used a differential pulse cathodic stripping voltammetry in an attempt to improve the sensitivity of the voltammetric determination of selenium after extraction as the 3,3′-diaminobenzidine piazselenol. After formation of the piazselenol, the selenium was deposited on a mercury electrode at −0.45 V. The limit of detection of selenium by this method was 0.01 μg litre^{-1}. Interferences could be avoided by extraction of the piazselenol into toluene followed by back-extraction into 0.5 M hydrochloric acid.

Donat and Bruland[34] preconcentrated cyclohexane-1,2-dione dioxime (nioxime) complexes of cobalt(II) and nickel(II) from 10 ml sea-water samples on to a hanging mercury drop electrode by controlled adsorption. Cobalt(II) and nickel(II) reduction currents were measured by differential pulse cathodic stripping voltammetry. Detection limits for cobalt and nickel were 6 pM and 0.45 mM, respectively.

7.1.2 Gas evolution methods

Hydride formation In this procedure, the element of interest is converted to a hydride which is then preconcentrated in a cold trap, then released for estimation by a suitable technique; usually atomic absorption spectrometry or gas chromatography.

A number of elements in the fourth, fifth, and sixth groups of the periodic system form hydrides upon reduction with sodium borohydride,[48] which are stable enough to be of use for chemical analysis (Ge, Sn, Pb, As, Sb, Se, Te). Of these elements, Andreae[35] has investigated in detail arsenic, antimony, germanium, and tin. The inorganic and organometallic hydrides are separated by a type of temperature-programmed gas chromatography. In most cases it is optimal to combine the functions of the cold trap and the chromatographic column in one device. The hydrides are quantified by a variety of detection systems which take into account the specific analytical chemical properties of the elements under investigation. For arsenic, excellent detection limits (about 40 pg) can be obtained with a quartz tube cuvette burner which is positioned in the beam of an atomic absorption spectrophotometer. For some of the methylarsines, similar sensitivity is available by an electron capture detector. The quartz-burner-atomic absorption spectrometry system has a detection limit of 90 pg for tin; for this element much lower limits (about 10 pg) are possible with a flame photometric detection system, which uses the extremely intense emission of the SnH molecule at 609.5 nm. The formation of GeO at the temperatures of the quartz tube furnace makes this device quite insensitive for the determination of germanium. Excellent detection limits (about 140 pg) can be reached for this element by the combination of the hydride generation system with modified graphite furnace atomic absorption spectrometry.

Many of the recent methods make use of the condensation of the hydrides in a cold trap at liquid nitrogen temperature. Braman and Foreback[49] pioneered the use of a packed cold trap to serve both as a substrate to collect the hydrides at liquid nitrogen temperature, and to separate arsine and the methylarsines gas chromatographically by controlled heating of the trap. In the same paper, they described the differentiation between As^{III} and As^{V} by a prereduction step and by control of the pH at which the reduction takes place. The lowest detection limits were achieved by cold-trapping of the hydrides and subsequent introduction into either a quartz cuvette furnace or into a commercial graphite furnace.[36]

The determination of the hydride element species consists of five steps:

(1) the reduction of the element species to the hydrides;

(2) the removal of interferent volatiles from the gas stream;

(3) the cold-trapping of the hydrides;

(4) the separation of the substituted and unsubstituted hydrides from each other and from interfering compounds; and

(5) the quantitative detection of the hydrides.

A typical instrumental configuration to accomplish these steps is shown in Fig. 15 for the borohydride reduction/flame photometric detection system for tin speciation analysis.

Only when the very contamination-sensitive electron-capture detector is used is it necessary to provide separate gas streams, one for the reaction and stripping part of the system, the other for the carrier gas stream of the the column and detector. Otherwise, the same gas stream can be used to strip the hydrides from solution and to carry them into the detector, which greatly simplifies the apparatus. Initially, column packings of glass beads or glass wool[37] were used; however, these packings produce poor separation of the methylated species from each other and badly tailing peaks. Andreae[35] therefore used a standard gas chromatographic packing (15 per cent OV-3 on Chromosorb W/AWDMCS, 60–80 mesh) in U-tubes for the separation of the inorganic and alkyl species of arsenic, antimony, and tin. This packing is quite insensitive to water and produces sharp and well-separated peaks.

Depending on the detector used, some volatile compounds which are formed or released during the hydride generation step may interfere with the detection of the hydrides of interest. Most prominent among them are water, carbon dioxide, and, in the case of anoxic water samples, hydrogen sulphide. The atomic absorption detector is insensitive toward

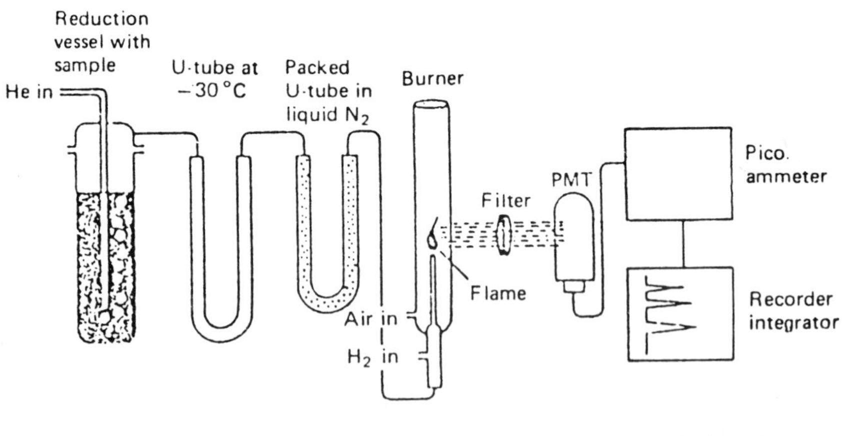

Fig. 15 Schematic design of the $NaBH_4$-reduction/flame photometric detection system for the determination of tin species in natural waters. From Braman and Foreback[49] with permission.

these compounds; thus no precautions need to be taken when this detector is used. It has been found convenient in some applications however, to remove most of the water before it enters the cold trap/column which serves to condense and separate the hydrides. This can be accomplished by passing the gas stream through a larger cold trap cooled by a dry ice/alcohol mixture or by an immersion cooling system.[38] This method was also used with water-sensitive detectors, e.g. the electron-capture detector for methylarsines[39] or with plasma discharge detectors (e.g. Crecelius[40]). ✓

Non-saline waters. Cutter[41] proposed a method for selenium in which various forms of selenium in a large volume of sample are reduced with sodium borohydride. The volatile selenium is swept off with helium and collected in a liquid nitrogen trap. The methyl selenium species are separated by gas chromatography and measured in a quartz tube atomic absorption furnace at levels corresponding to 5 ng litre^{-1} in the original sample. Other workers have applied a similar technique.[42,43]

Orvini *et al.*[44] determined arsenic(III) arsenic(V), and total inorganic arsenic using selective hydride generation coupled with neutron activation analysis. Shaikh and Tallman[45] determined arsenic with nanogram sensitivity by reducing arsenic species to hydrides and collecting them in a liquid nitrogen trap. They are selectively vaporized from the trap and directly injected into a graphite furnace atomic absorption spectrometry.

Hanson *et al.*[46] generated the gaseous hydrides of antimony(III), arsenic(III), and tin(II) by sodium hydride reduction. The hydrides were swept from a large volume of sample on to a Poropak column where they were separated and detected at a gold gas-porous electrode by measurement of their electro-oxidation currents. Detection limits were 0.2 μg litre^{-1} for arsenic(III) and antimony(III) and 0.8 μg litre^{-1} for tin(II).

Vien and Fry[47] employed a hydride generation gas chromatographic system with an intermediate cold trap that required no drying agents or carbon dioxide scrubbers in their preconcentration of arsenic, selenium, tin, and antimony from natural waters. The combination of a photoionization detection and cryogenic entrapment was also tolerant to water vapour, volatile boranes, and other spurious byproducts from a borohydride reactor, besides improving sensitivity by several orders of magnitude. Detection limits for arsenic, selenium, antimony, and tin were as low as 0.025 ng in a 28 ml sample.

Sea-water. Amaukwah and Fashing[36] have discussed the determination of arsenic(V) and arsenic(III) in sea-water by solvent extraction–absorption spectrometry using the hydride generation technique. Bertine and Lee[50] have described hydride generation techniques for determining Sb^{V} and Sb^{III}, Sb-S species, and organoantimony species in frozen sea-water samples.

The full potential of the hydride method has not yet been realized. Cutter[51] has studied its application to the determination of selenium in sea-water. Other elements which might be amenable to the technique are bismuth, tellurium, and lead. Willie *et al.*[52] used the hydride generation–graphite furnace atomic absorption spectrometry technique to determine selenium in sea-water. A Pyrex cell was used to generate selenium hydride which was carried to a quartz tube and then to a preheated furnace operated at 400 °C. Pyrolytic graphite tubes were used. Selenium could be determined in sea-water down to 20 ng $litre^{-1}$. No interference was found due to iron, copper, nickel, and arsenic.

Sturgeon *et al.*[53] have described a hydride generation atomic absorption spectrometric method for the determination of antimony in sea-water. The method uses formation of stibene using sodium borohydride. Stibine gas was trapped on the surface of a pyrolytic graphite coated tube at 250 °C and antimony determined by atomic absorption spectrometry. An absolute detection limit of 0.2 ng was obtained and a concentration detection limit of 0.04 μg $litre^{-1}$ obtained for 5 ml sample volumes.

Earlier investigations have shown that the major known organic arsenic compound in the enviroment is dimethyl arsinate. For that reason, specific and sensitive methods for the determination of this compound are needed. Several months exist for this determination often together with other arsenic species. Hydride generation and selective vaporization of cold-trapped arsines in combination with various detection systems seem to be the methods most frequently used.[54–56] These methods are applicable to the species $As^{(III)}$, As^{V}, monomethyl arsinates, dimethyl arsinate and trimethylarsine oxide. The optimum conditions for generation of the respective arsine are, however, different for all these species with regard to the pH of the generating solution.[57,58] These methods also suffer from interferences from numerous inorganic ions.[59] Preconcentration in a toluene cold-trap following arsine generation was successful for monomethyl arsinate and dimethyl arsinate but non-quantitative recoveries of dimethyl arsinate were reported, probably because of the problem mentioned above with the arsine generation step. Molecular rearrangements occurring during arsine generations are, however, reported to be minimized if sodium tetrahydroborate is introduced as a pellet.[60] To simplify the determination of the different species, improved separation of the arsines using gas chromatography may be necessary.

The simplest analytical method is direct measurement of arsenic in volatile methylated arsenicals by absorption.[61] A slightly more complicated system, but one that permits its differentiation of the various forms of arsenic, uses reduction of the arsenic compounds to their respective arsines by treatment with sodium borohydride. The arsines are collected in a cold trap (liquid nitrogen) then vaporized separately by slow warming, and the arsenic measured by monitoring the intensity of an arsenic spectral line, as produced by a direct current electrical discharge.[62,63,65] Essentially the same method was proposed by Talmi and Bostick[64] except that they collected the arsines in cold toluene (-5 °C), separated them on a gas chromatography column and used a mass spectrometer as the detector. Their method had a sensitivity of 0.25 μg litre^{-1} for water samples. Another variation on the method[66] with slightly higher sensitivity (several nanograms per litre) used the liquid nitrogent cold trap and gas chromatography separation, but used the standard gas chromatography detectors or atomic absorption, for the final measurement. These workers found four arsenic species in natural waters.

Burton and co-workers[67] have studied the distribution of arsenic in the Atlantic Ocean. Samples from 1000 m and above were filtered through acid-washed 0.45 μm Sartorius membrane filters. Analyses on samples from depths below 1000 m were made on unfiltered water. Aliquots of 50 ml were placed in a round-bottomed flask, fitted with a modified Dreschel head and an injection syringe in a side arm. Concentrated hydrochloric acid (20 ml), 1 ml of 1 M ascorbic acid solution, and 1 ml 1 M potassium iodide solution were added. The solution was stood for 30 min to allow reduction of As^{V} to As^{III}, which was necessary to ensure quantitative recovery of inorganic arsenic as arsine under the conditions used in the subsequent step. With nitrogen passing through the flask at a flow rate of 150 ml min^{-1}, 0.5 ml 8 per cent w/v sodium borohydride solution was added from the syringe. The arsine evolved was trapped in 2 ml of a solution containing 0.7 per cent w/v potassium iodide, and excess iodine, over a period of 3 min.

The concentrates were subsequently analysed for arsenic using a Varian-Techtron AA5 atomic absorption spectrophotometer fitted with a Perkin-Elmer HGA 72 carbon furnace, linked to a zinc reductor column for the generation of arsine. A continuous stream of argon was allowed to flow with the column connected into the inert gas line between the HGA 72 control unit and the inlet to the furnace. Calcium sulphate (10–20 mesh) was used as an adsorbent to prevent water vapour entering the carbon furnace. The carbon tube used was of 10 mm i.d. and had a single centrally located inlet hole.

A wide range of elements was tested for interfering effects; the only significant interferences found were at concentrations much higher than those encountered in sea-water. No significant difference in the results was found when a sample of sea-water was analysed in the way described and also by the same procedure but using the method of standard additions.

Carbonyl formation

Sea-water. Lee[68] described a method for the preconcentration of nanogram or subnanogram amounts of nickel in sea-water. Dissolved nickel is reduced by sodium borohydride to its elemental form which combines with carbon monoxide to form nickel carbonyl. The nickel carbonyl is stripped from solution by a helium–carbon monoxide mixed gas stream, collected in a liquid nitrogen trap, and atomized in a quartz tube burner of an atomic absorption spectrophotometer. The sensitivity of the method is 0.05 ng of nickel. The precision for 3 ng nickel is about 4 per cent. No interference by other elements is encountered in this technique. Between 0.3 and 0.6 μg litre^{-1} nickel was found by this method, in a vertical profile of water samples taken down to 1200 m in the Santa Catalina Basin.

7.1.3 Flotation methods

Co-flotation

Non-saline waters. Hidalgo *et al.*[69] preconcentrated molybderium (VI) in potable water and sea-water by co-flotation on iron(III) hydroxide. Co-flotation was achieved by means of surfactants; octadecylamine for tap water samples, hexadecyltrimethylammonium bromide for natural waters, and both hexadecyltrimethylammonium bromide and octadecylamine for sea-water samples. Differential pulse polarography using the catalytic wave caused by molybdenum(VI) in nitrate medium was applied to the preconcentrate. Good reproducibility was obtained with mean values of 0.7 ng ml^{-1} and 5.7 ng ml^{-1} for molybdenum(VI) in tap water and sea-water, respectively.

Sea-water. Hiraide *et al.*[72] used continuous flow coprecipitation–flotation for the radiochemical separation of cobalt-60 from sea-water. The cobalt-60 activity was measured by liquid scintillation counting with greater than 90 per cent yield and a detection limit of 5 fCi litre^{-1} sea-water.

Co-flotation with octadecylamine and ferric hydroxide as collectors has been used to separate copper, cadmium and cobalt from sea-water.[73] The method was based on the co-flotation or adsorbing colloid flotation technique. The substrates were dissolved in an acidified mixture of ethanol, water, and methyl isobutyl ketone to increase the sensitivity of the determination of these elements by flame atomic absorption spectrophotometry. The results were compared with those of the usual ammonium pyrrolidinedithiocarbamate–methyl isobutyl ketone extraction method. While the mean recoveries were lower, they were nevertheless considered satisfactory (Table 55).

Table 55 Cu^{II}, Cd^{II}, and Co^{II} recovery from spiked sea-water samples by the co-flotation method (V = 2000 ml) and the APDC/MIBK method (V = 400 ml).

Method	Element addition (ng ml^{-1})	Species (ng ml^{-1})					
		Cu^{II}		Cd^{II}		Co^{II}	
		found	RSD (per cent)	found	RSD (per cent)	found	RSD (per cent)
Co-flotation	0.0	2.5	13.3	0.8	6.0	5.8	7.0
	5.0	5.6	9.5	4.3	6.0	11.4	5.6
	10.0	10.1	2.0	8.4	3.4	15.9	4.7
	Mean recovery (per cent)	83 ± 6		79 ± 4		98 ± 5	
APDC/MIBK	0.0	4.1	29.4	2.8	14.5	3.2	22.0
	5.0	7.0	4.4	7.6	7.2	7.5	2.7
	10.0	13.8	8.2	12.0	4.9	11.7	1.7
	Mean recovery (per cent)	87 ± 6		94 ± 2		92 ± 2	

From Cabezon *et al.*[73] with permission.

Froth flotation

Sea-water. Tseng and Zeitlin[70] studied the applicability of an intrinsically rapid technique namely adsorption colloid flotation. This separation procedure uses a surfactant-collector–inert gas system, in which a charged surface-inactive species is adsorbed on a hydrophobic colloid collector of opposite charge; the colloid with the adsorbed species is floated to the surface with a suitable surfactant and inert gas, and the foam layer is removed manually for analysis by a methylene blue spectrometric procedure. The advantages of the method include a rapid separation, simple equipment and excellent recoveries. These workers used the flotation unit that was devised by Kim and Zeitlin.[71] The recovery of the selenium was found to be 100 ± 10 per cent (at the 95 per cent confidence level). Using this method Tseng and Zeitlin[70] found 0.40 ± 0.12 μg litre^{-1} selenium in sea-water.

Colloid flotation

Sea-water. Adsorbing colloid flotation has been used to separate uranium from sea-water.[75] To the filtered sea-water (500 ml; about 1.5 μg U) is added 0.05 M ferric chloride (3 ml), the pH is adjusted to 6.7 ± 0.1 and the uranium present as $(UO_2(CO_3)_3)^{4-}$ is adsorbed on the colloidal ferric hydroxide which is floated to the surface as a stable froth by the addition of 0.05 per cent ethanolic sodium dodecyl sulphate (2 ml) with an air-flow (about 10 ml min^{-1}) through the mixture for 5 min. The froth is removed and dissolved in 12 M hydrochloric acid – 16 M nitric acid (4:1) and the uranium is salted out with a solution of calcium nitrate containing EDTA, and determined spectrophotometrically at 555 nm by a modification of a Rhodamine B method. The average recovery of uranium is 82 per cent; co-adsorbed WO_4^{2-} and MoO_4^{2-} do not interfere.

Adsorbing colloid flotation has also been used by Williams and Gillam.[76] The fusion track method has also been used by Hashimoto.[74] In this method, the uranium is first coprecipitated with aluminium phosphate,[77] the precipitate is dissolved in dilute nitric acid, and an aliquot of the solution is transferred to a silica ampoule into which small pieces of muscovite are inserted before sealing. The uranium is then determined by measuring the density of fission tracks formed on the muscovite during irradiation of the ampoule for 15 min at 80 °C in a reactor. The muscovite is etched with hydrofluoric acid for 1 h before the photomicrography; the density is referred to that obtained with standard solution of uranium. There is no interference from thorium, and no cheimical separations are required. An average concentration of 3–40 ± 12 μg uranium per litre was obtained, in good agreement with the normally accepted value.

Voyce and Zeitlin[78] have used adsorption colloid flotation to determine mercury in sea-water. The sample (500 ml) is treated with concentrated hydrochloric acid, an aqueous solution of cadmium sulphate and a fresh aqueous solution of sodium sulphate are added. The pH is adjusted to pH 1.0 and solute poured into a flotation cell with a nitrogen flow of 10 ml min^{-1}. Ethanolic octadecyltrimethylammonium chloride is injected and the froth dissolved in aqua regia in a flameless atomic absorption cell. Following the reduction of mercury with stannous chloride the mercury vapour is flushed from the system. To determine organically bound mercury, the sample (500 ml) is treated with 0.5 M sulphuric acid and aqueous potassium permanganate and set aside for 24 h. Aqueous hydroxylammonium chloride is added and the determination completed as above. The amounts of mercury in the samples are calculated by reference to the standard absorptions. Average recoveries of 0.05 μg mercury were 88 per cent.

Murthey and Ryan[79] used colloid flotation as a means of preconcentration prior to neutron activation analysis for arsenic, molybdenum, and vanadium. Hydrous iron(III) is floated in the presence of sodium dodecyl sulphate with small nitrogen bubbles from 1 litre of sea-water at pH 5.7. Recoveries of arsenic, molybdenum, and vanadium were better than 95 per cent whilst that of uranium was about 75 per cent. Adsorption colloid flotation using dodecylamine as surfactant has been used[80] to separate zinc with 95 per cent efficiency from sea-water.

7.1.4 *Evaporation/distillation methods*

Solvent sublation

Non-saline waters. Solvent sublation is a flotation process in which the material adsorbed on the surface of gas bubbles is collected on a layer of immiscible liquid instead of as a layer of foam over an aqueous phase as occurs in the conventional foam flotation process.

Cervera *et al.*[81] studied the feasibility of using this process for the separation and preliminary concentration of copper as its dithizonate. Sodium lauryl (sulphate) was used as the collector with dithizone solutions in methyl isobutyl ketone. Cervera *et al.*[81] established optimum experimental conditions. Sample volumes of several litres can be handled making this a very effective process with the possibility of preconcentration factors of up to 100. The experimental conditions are described below:

Reagents:

All reagents and solvents were of analytical reagent grade and were used without further purification. The solutions were diluted as necessary to prepare working standard solutions.

Copper stock solution, 1000 mg litre^{-1}.

Sodium lauryl sulphate, standard solution, 0.01 M.

Dithizone in isobutyl methyl ketone stock solution, 0.324 g litre^{-1}.

Apparatus:

The experimental device used is shown in Fig. 16. A water cooled high efficiency reflux condenser was attached on the upper part of the flotation column to prevent evaporation losses from the upper organic layer, which was important in long-duration experiments (more than 1 h). The column had a capacity of about 3.5 litres and a cross-section of 70 cm^2 and was fitted at the lower end with a detachable section furnished with a sintered glass place of low porosity (no. 3). A Pye Unicam SP 9800 atomic absorption spectrophotomoter using an air–acetylene flame or equivalent was used for determining concentrations of copper(II) in aqueous layer during sublation process.

An air stream was used as the carrier gas, presaturated with water vapour, using a membrane pump and the flow was regulated by means of a needle valve. The gas flow rate was controlled by means of a bubble rotameter. An optimum flow rate of carrier gas (air pre-saturated with water vapour to minimize carriage of vapour from the aqueous phase by the ascending gas bubbles) at a limit of 80 ml min^{-1} was adopted. The optimum flow rate of the carrier gas depends on a number of parameters

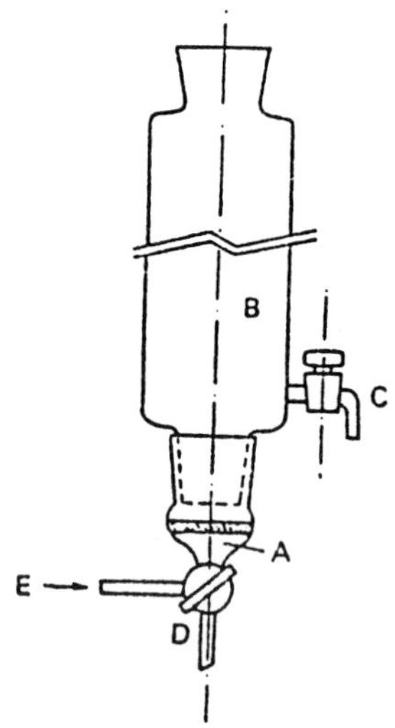

Fig. 16 Experimental device used for the solvent sublation process. A, Detachable section with sintered glass plate; B, variable size column; C, stopcock; D, three way stopcock; and E, gas inlet. From Cervera *et al.*[81] with permission.

directly related to the porosity and mechanical quality of the sintered glass plate used, to the column size (especially its cross-section), and to the salinity of the aqueous phase (an increase of which favours mixing of the two phases). Care should be taken to keep the flow rate reasonably constant, which is fairly easy as the samples to be taken from the aqueous phase during the process are small enough not to bring about a significant variation in flow rate owing to the resulting charge losses of the liquid column, which produce slight increases in the flow rate.

Under the experimental conditions used, the total decrease in height of the aqueous phase column arising froin the periodic extraction of samples for analysis did not exceed about 3 per cent the initial height thus making it unnecessary to make flow corrections following removal of samples for analytical monitoring.

Attention must be paid however, to the eventual transport of organic phase by the gas bubbles, which might reach a considerable extent owing to the small volume of organic phase used (25 ml) especially when the process is run for long periods (up to several hours). This problem can be greatly reduced by attaching a suitable water cooled glass condenser of high efficiency to the upper end of the column combined with the use of organic solvents of low volatility (methyl isobutyl ketone, octan-2-ol, etc). Provided that the experiments are carried out within a reasonable time (about 1 h or less) there is no need to take special precautions with the dithizone solutions, although it is easy to isolate the system from heat and light.

Procedure:

Add suitable amounts of 0.01 M sodium lauryl sulphate solution and ethanol to 1 litre of sample and adjust the pH to 5.9 to 6.1 with hydrochloric acid or sodium hydroxide solution. Upon transfer of the adjusted solutions into the flotation column, add 25 ml of dithizone in methyl isobutyl ketone solution to the aqueous layer (avoiding mixing of the phases) before the solvent sublation process started. Take samples of the aqueous phase periodically for analytical monitoring of the process by means of the stopcock attached to the lower part of the flotation device (Fig. 16(c)). Upon acidification, when needed, inject the samples in an atomic absorption spectrophotometer to determine the residual copper(II) concentration in the aqueous phase. Use calibration graphs prepared with standard copper samples under the same experimental conditions (detergent concentration, ethanol, pH, salinity).

Cervera *et al.*[81] demonstrated that their solvent sublation process was not applicable to the determination of trace metals in estuarine and seawater. They concluded that the solvent sublation technique allows a highly effective separation and preconcentration of trace amounts of

Table 56 Comparative results of copper(II): dithizone solvent sublation and solvent extraction obtained under the same experimental conditions.

Experiment No.	NaLS:Cu^{II} molar ratio	Solvent extraction		Solvent sublation	
		Extraction (per cent)	Time to reach equilibrium (min)	Yield (per cent)	Time necessary (min)
1	40:1	73.3	~30	100	~20
2	15:1	75.6	~20	95.6	~45
3	7:1	78.4	<3	78.3	—[a]
4	1:1	78.2	<2	28.4	—[a]
5	0:1	78.5	<1	—	—

[a] Yield obtained after 60 min.
From Cervera *et al.*[81] with permission.

copper from aqueous solution as the dithizonate within a narrow pH range. The yield of the process is 100 per cent reaching this value for sodium lauryl sulphate to copper(II) ratios greater than 20:1 in not more than 20 min.

In order to obtain more accurate results for comparative purposes, a separate extraction process was carried out using 250 ml extraction funnels and keeping the same organic to aqueous layer volume ratios used in the solvent sublation method. The kinetics and yields of the corresponding processes (using in all instances an initial copper(II) concentration of 1 mg litre^{-1}) are compared in Table 56, which demonstrates clearly the difference between the two processes and the outstanding concentration features exhibited by the solvent sublation process.

Direct evaporation

Non-saline waters. Although this is an obvious method for preconcentrating water samples prior to analysis, this technique has several disadvantages, namely the risk of sample contamination and the lengthy nature of the evaporative process. Nevertheless some work has been carried out on the application of evaporative methods[88-90] to the determination of trace elements[89] including copper[90] Preconcentration of dilute solutions of 12 metals at the 10^{-12} g litre^{-1} level by non-boiling evaporation followed by atomic absorption spectropbotometry has been applied.[89]

Freeze drying

Non-saline waters. Crocker and Merritt[82] preconcentrated water by freeze drying prior to the determination of 0.1 μg litre^{-1} amounts of elements of atomic weight 7 to 238 in rain-water by mass source mass spectrometry. Miscellaneous elements (30) in lake water have been preconcentrated by freeze drying prior to determination at the 0.01 μg litre^{-1} level by spark source mass spectrometry.[83]

Lieser *et al.*[84] used freeze drying as a means of preconcentrating potable and lake water samples prior to the determination of trace metals by neutron activation analysis of the dried solid. Some results obtained by this procedure are given in Table 57. These show the sensitivity, reproducibility, and accuracy of the procedure. Unfortunately, the authors give no details of the freeze drying procedure they adopted.

To preconcentrate ground water samples prior to the determination of uranium-239 by epithermal neutron activation analysis Juleli and Parry[85] freeze dried 50 ml aliquots of filtered and acidified ground water samples. Uranium-239 peaks of gamma-ray energy at 74.7 keV were measured to calculate uranium concentrations applying the total peak area method against that of a uranium standard. Results of this

Table 57 Relative mean values and standard deviations obtained by freeze drying.

Element	Concentration (μg litre^{-1})	Relative mean value and standard deviation, all values included
As	1	1.81 ± 2.4
Au	0.01	0.36 ± 0.20
Ba	100	0.67 ± 0.037
Br	10	1.08 ± 0.14
Ca	10^5	1.04 ± 0.17
Cd	1	4.11 ± 3.12
Ce	0.1	4.84 ± 0.91
Co	1	1.10 ± 0.11
Cr	1	1.04 ± 0.34
Cu	10	2.75 ± 3.0
Eu	0.1	1.21 ± 0.55
Fe	100	0.72 ± 0.35
Hg	0.1	31.2 ± 106
K	1000	1.16 ± 0.58
La	0.1	1.14 ± 0.32
Mo	1	1.08 ± 0.26
Na	5000	0.87 ± 0.09
Sb	1	0.90 ± 0.33
Sc	0.02	1.29 ± 0.18
Se	1	1.04 ± 0.20
U	1	1.17 ± 0.15
Zn	10	0.98 ± 0.09

From Lieser *et al.*[84] with permission.

technique were compared against a set of data obtained by conventional instrumental neutron activation using neptunium-239 and a set of data from laser induced fluorimetry (LIF). The three data sets obtained for ground water samples and for US Geological Survey standard reference samples were in good agreement especially when the total dissolved solids of the samples was low.

Sea-water. Gordon and Larson[86] used photon activation analysis to determine strontium-87 in sea-water. Samples (2 ml, acidified to pH 1.67 or 2.54 for storage) were filtered and freeze dried. The residues, together with strontium standards, were wrapped in polyethylene and aluminium foil and irradiated in a 30 MeV bremsstrahlung flux of γ-radiation. After irradiation, the samples were dissolved in 50 ml of acidified water and ^{87m}Sr was separated by precipitation as strontium carbonate for counting (γ-ray spectrometer, Ge(Li) detector, and multi-channel pulse-height analyser). The standard deviation at the 7 ppm strontium level was ± 0.47. Freeze drying followed by neutron activation analysis enabled Jørstad and Selbu[87] to determine to 0.7 μg litre^{-1} selenium in sea-water.

7.1.5 Precipitation and cocrystallization methods

Cocrystallization

Non-saline waters. This technique has been subject to a limited amount of investigation in the case of molybdenum organic chelate complexes.[91–94]

Reductive precipitation

Non-saline waters. Skogerboe *et al.*[95] preconcentrated trace elements from water samples by precipitation as the reduced element or boride. Trace iron was added as a carrier to improve recovery.

7.1.6 Dialysis and osmosis

Dialysis

Non-saline waters. Dialysis has been applied to the determination the chemical form of trace elements in natural water.[96,97] Benes and Steinnes[97] have described a method in which the dialysis bag filled with pure water is immersed in the natural water sample, allowing the dialysis and adsorption equilibrium to be established.

Morrison[136] estimated bioavailable metal uptake rates for storm water and treated sewage effluents using dialysis with receiving resins. The metal chelating resin, Chelex-100, in the calcium form was incorporated within a dialysis bag sealed at both ends. The bag was immersed in the sampled solution for 1–4 days before releasing the chelating resin to a separating column. Metals were eluted front the resin with 1 M nitric acid followed by ultra-pure water and concentrations determined by graphite furnace and flame atomic absorption spectrometry. Results were expressed in terms of a metal uptake rate per unit surface area and time (pg $mm^{-2}.h.^{-1}$). Background resin uptake rates from an unpolluted stretch of river for zinc (0.79), cadmium (0.01), lead (0.02), and copper (0.06) were exceeded at three storm-water outfalls (Backebolsimotet, Bergsjon, and Floda) which provided the following range of resin uptake rates for zinc (2.5–67.1), cadmium (0.01–0.14), lead (0.15–2.4), and copper (0.26–41.3). Mean efficiencies for the reduction of bioavailable uptake rates between influent and effluent water of a sewage treatment plant were derived for zinc (42 per cent), lead (49 per cent), and copper (33 per cent).

Hart and Davies[96] combined dialysis with ion exchange and this combination enabled the determination to be completed in a much shorter time (5 h) than in the conventional dialysis methods (24 h). They showed that about 100 per cent of cadmium, copper, lead, zinc, and iron were dialysed in the 5 h period. The results gave information on the concentration of ionic species present in the natural water sample. The experimental set up is illustrated in Fig. 17. The dialysis unit was from a Technicon Auto Analyzer II: 'Cuprophanel' membranes were used. This membrane consists of regenerated cellulose with a pore size of approximately

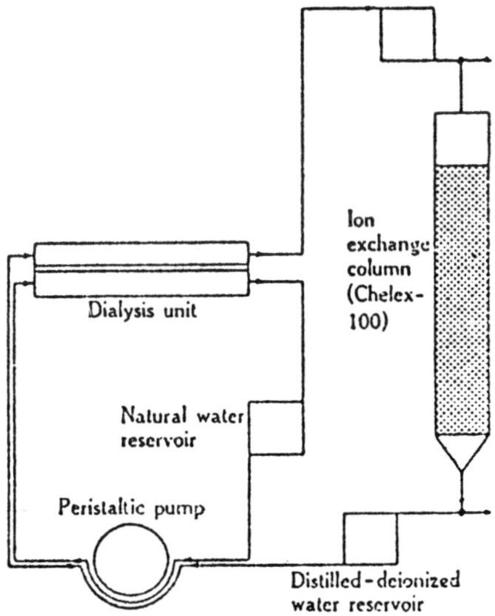

Fig. 17 Dialysis ion exchange apparatus. From Hart and Davies[96] with permission.

0.8 nm. The length of travel of liquids through the dialyser was approximately 250 cm.

Two peristaltic pumps (Peripump type 82–334) were used to force liquid through the dialysis unit at a rate of approximately 5 ml min^{-1}. An ion exchange column containing 3 g of Chelex-100 resin (in sodium form) was used to remove ions from solution. The column and reservoirs were made from borosolicate glass and were connected with 5 mm o.d. Tygon tubing. New membranes were soaked for 30 min in 1 M nitric acid and rinsed with double distilled water that had also been passed through a Milli-Q2 system. All other apparatus in contact with samples was thoroughly washed in a 2 per cent solution of Extran (Merck) soaked in 6 M nitric acid for at least 24 h, and finally rinsed with double distilled deionized water. The dialyser was cleaned between experiments by circulating 1 M nitric acid through both donor and receptor channels for approximately 30 min, followed by thorough rinsing with double distilled deionized water (4 volumes of 5 ml).

The Chelex 100 resin (Bio Rad Labs, 100–200 mesh) was decontamininated before use by washing with 5 ml of 1 M nitric acid (BDH Aristar grade) and then converted into the sodium form by soaking in 2 M sodium acetate (Merck Suprapur grade). The resin was then placed in the column and washed thoroughly with four 5 ml volumes of distilled deionized water. Nitric acid (1 M) as used to elute the ion exchange column.

Analyses were mainly performed using atomic absorption spectrophotometry—Varian Techtron AA with a laminar flow burner (Techtron type AB 51) or carbon rod furnace (CRA-63). In Table 58 are shown the satisfactory transition metal recoveries obtained following a 5 h dialysis of a synthetic natural water sample containing typical levels of sodium, potassium, calcium, magnesium, and anions expected in natural water samples.

Table 58 Dialysis recovery experiments using synthetic natural waters.[a]

	Trace metal				
	Cd	Cu	Pb	Zn	Fe
Initial conc. μg $litre^{-1}$	1.0	6.25	6.25	10.0	100.0
Recovery[b] (per cent)	104 ± 6	98 ± 5	95 ± 5	100 ± 2.5	103 ± 4

[a] Synthetic natural waters contained the following (in μg $litre^{-1}$): Na 4.72, K 1.02, Ca 3.61, Mg 1.73, HCO_3 27.45, Cl 2.91, SO_4 0.96, and the pH was 6.9. Percentage recovery values are means ± s.e.m. calculated from the results of five determinations on each of two dialysis experiments.
[b] Analysis methods: Cd, Cu, Pb and Zn by AA-C rod; Fe by AA-flame.
From Hart and Davies[96] with permission.

In Table 59 are shown full analyse obtained by this technique of three river samples. Water samples were taken using an acid-washed Van Dorn water sampler. Samples for total trace metal determination were acidified on site with concentrated nitric acid (5.0 ml $litre^{-1}$) and stored in 1-litre acid washed polythene bottles. The other samples were taken back to the laboratory for filtration and further processing. Filtration was achieved in a Perspex pressure filtration apparatus using acid-washed 0.4 μm Nuclepore membrane filter. The time elapsing between sampling and the completion of filtration was approximately 2 h. The filtered sample was split into two portions. One portion of 200 ml was acidified to pH 1–2 with concentrated nitric acid, boiled gently for 30 min, and analysed by anodic scanning voltammetry to give total filterable trace metal concentrations. A second portion of 100 ml was dialysed against 25 ml of double distilled deionized water for 5 h, using the experimental technique previously described. The column was

Table 59 Analyses of natural water samples, values expressed are means ± s.c.m.

Sample No.[a]	Element	Trace metal concentration (μg $litre^{-1}$)			
		Total[b]	Total filterable[b]	Dialysable ionic[c]	Dialysable non-ionic[d]
1	Cd	0.50 ± 0.07	0.37 ± 0.05	0.42 ± 0.01	<0.02
	Cu	3.55 ± 0.14	3.0 ± 0.47	1.47 ± 0.12	<0.2
	Pb	34.5 ± 2.1	3.15 ± 0.48	2.36 ± 0.10	<0.2
	Zn	84 ± 2	50 ± 2	17.0 ± 1.3	<0.5
	Fe	570 ± 20	150 ± 20	< 20	<50
2	Cd	0.30 ± 0.005	0.21 ± 0.02	0.24 ± 0.01	<0.02
	Cu	6.3 ± 0.23	5.2 ± 0.97	1.5 ± 0.12	<0.2
	Pb	10.0 ± 1.8	6.7 ± 0.5	2.1 ± 0.1	<0.2
	Zn	71 ± 2	55 ± 2	14.8 ± 0.2	<0.5
	Fe	690 ± 20	620 ± 20	<20	<50
3	Cd	0.35 ± 0.01	0.30 ± 0.01	0.22 ± 0.005	<0.02
	Cu	5.5 ± 0.35	3.5 ± 0.22	3.0 ± 0.2	<0.2
	Pb	21 ± 1.7	10.1 ± 0.5	2.3 ± 0.1	<0.2
	Zn	45 ± 2	23 ± 2	5.7 ± 0.14	<0.5
	Fe	850 ± 25	570 ± 25	<20	<50

[a] Sample 1: Yarra River at Hoddle Bridge, 8 October 1975. Sample 2: Yarra River at Hoddle Bridge, 28 January 1976. Sample 3: Dandenong Creek at Ferntree Gully Road, 29 January 1976.
[b] Analysis methods: Cd, Cu, and Pb by ASV; Zn and Fe by AA-flame.
[c] Analysis methods: Cd, Cu, Pb, and Zn by AA-C rod: Fe by AA-flame.
[d] Analysis methods: Cd, Cu, Pb, and Zn by ASV; Fe by AA-flame.
From Hart and Davies[96] with permission.

eluted, and the eluate analysed with atomic absorption carbon rod to give dialysable ion exchangeable trace metal concentrations. The dialysate remaining after dialysis should contain any trace metal that was dialysed but was not retained on the Chelex-100 column.

The results show that basically all the filterable cadmium exists in a dialysable form in all three samples. The proportion of zinc in a dialysable form is considerably less than for cadmium. The ratio of dialysable to filterable forms is quite variable for copper and lead over the three samples. This variability is no doubt due to the differences in water chemistry between the three samples. As expected, there is very little iron in a dialysable form in any of the sampler. Presumably most of the filterable iron is in the form of very fine colloidal hydroxyoxide species. In the three samples studied, the results show that cadmium, copper, and zinc exist predominantly in a form which is less than 0.4 μm in diameter (i.e. these trace metals all have a larger total filterable form than particulate form). The variability of the results for lead do not allow any conclusions to be drawn.

Wilson and Di Nunzio[98] used dialysis to preconcentrate nickel and cobalt by factors of up to 20 at the 200 μg litre^{-1} level.

Osmosis

Non-saline waters. Stec *et al.* have investigated the preconcentration of transition metals ions from dilute aqueous solution using osmosis.[99] They describe the manufacture, separation, and performance characteristics of the osmotic preconcentration cell. Cellulose acetate membranes were used as semi-permeable barriers between sample solutions and saturated salt solutions. These membranes were composed of dense mircoporous surface layers with highly porous substructure. After assembly of the cell, saturated salt solutions and water were cycled through the outer and inner channels, respectively, for approximately 15 min to assure a steady state permeation rate. Analyte recovery data were obtained on a multielement inductively coupled plasma emission spectrometer. The stability of permeation rates was dependent on the three parameters: temperature, salt concentration, and the flow rate of salt solution through the cell. Complexation of metal ions with EDTA prior to osmotic preconcentration was used to prevent loss of analytes by permeation through the membrane. A comparison of chelated and nonchelated recoveries for cadmium, copper, manganese, nickel, and zinc was made and the results tabulated. The reproducibility for unchelated ion recoveries was variable (95–64 per cent) but the major limitation of this technique was the back-permeation of sodium chloride from the saturated salt solution.

7.2 Anions

7.2.1 Ion flotation

Non-saline waters Aoyama *et al.*[100] have described a rapid method for the determination of down to 3 μg litre^{-1} of chromate in which hexavalent chromium is complexed with diphenylcarbazide, and this complex is floated with an sodium lauryl sulphate anionic surfactant. After dilution the subsided foam is measured spectrophotometrically. The continuous flotation procedure is as follows.

Place the water (1 litre), previously acidified to 0.1 M with sulphuric acid, in the separation tube, pass nitrogen into the tube at about 105 ml min^{-1}, and pump in the 1 per cent sodium lauryl sulphate solution at 0.39 ml min^{-1}. Continuously supply the sample ph and the 1 per cent diphenylcarbazide solution to the separation tube at a rate of 2 litre h^{-1} and 0.230 ml min^{-1} respectively. A steady state flotation was achieved 1 h after flotation had started. Mix the sample and diphenyl carbazide solutions in the mixing chambers before entering the tube. Collect the foam subsided in the collector and intermittently transfer to flasks and dilute to 50 ml with water, and measure absorbances as described above.

The effect of diverse ions on the flotation of chromium(VI) were investigated. Only copper(II) and iron(III) in 10-fold amounts, and vanadium(V) in 20-fold amounts caused fading of the colour. Even in those cases, however, chromium(VI) could be measured by a standard addition method.

In flotation has also been applied to the preconcentration of phosphate.[101]

7.3 Organics

7.3.1 Electrolytic methods

Electrlytic stripping

Sea-water. Wasik[102] has used an electrolytic stripping cell to determine hydrocarbons in sea-water. Dissolved hydrocarbons in a known quantity of sea-water were equilibrated with hydrogen bubbles, evolved electrolytically from a gold electrode, rising through a cylindrical cell. In an upper head-space compartment of the cell, the hydrocarbon concentration is determined by gas chromatography. The major advantages of this cell are that the hydrocarbons in the upper compartment are in equilibrium with the hydrocarbons in solution and that the hydrogen used as an extracting solvent does not introduce impurities into samples.

Wasik[102] used this method to successfully determine μg litre^{-1} of gasoline in sea-water. He found that a convenient method for concentration of the hydrocarbons is to recycle the hydrogen stream containing the hydrocarbons many times over a small amount of charcoal (2.3 mg). The charcoal, while still in the filter tube, was extracted three times with 5 μl carbon disulphide. A 2–3 μl aliquot of this solution was then injected into a SCOT capillary column.

A 10 ml portion of the sea-water extracted was diluted to 1000 ml with sea-water. To this solution was added toluene-*d*, as an internal standard to give a concentration of 0.05 ppm. The diluted extract wa.s poured into a 1000 ml stripping cell located in a constant temperature water bath. The temperature of the bath was raised to 80 °C and a current of 0.3A was passed through the cell. The upper chromatogram shown in Fig. 18 was obtained by sampling 1 ml of the head space after 30 ml hydrogen had bubbled through the stripping cell. The chromatogram shows the large number of aliphatic and olefinic hydrocarbon peaks that are eluted at the same time as the early aromatic hydrocarbon peaks. The lower chromatogram in Fig. 18 was obtained after 46 ml (V/V_L = 0.046) of hydrogen

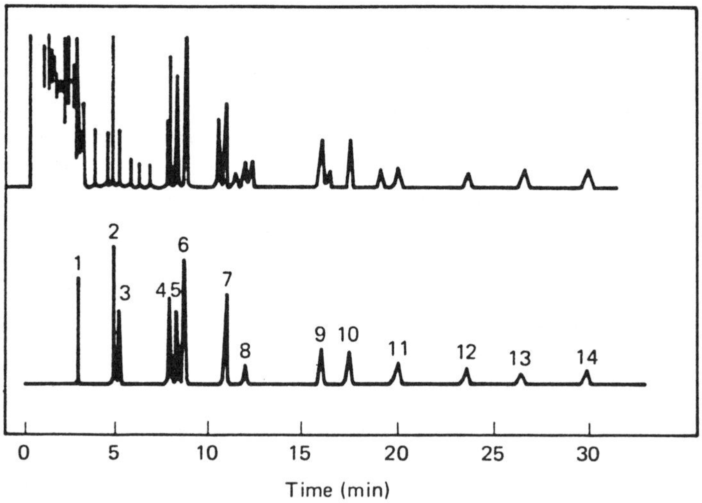

Fig. 18 Hydrocarbons stripped from artificial sea-water after 10 ml hydrogen had bubbled through the stripping cell (upper chromatogram) and after 100 ml hydrogen had bubbled through the stripping cell (lower chromatogram). A 1.0 ml volume of head-space gas was injected into a Scot column, 15 m × 0.5 mm i.d., coated with *m*-bis(*m*-phenoxyphenoxy)benzene; carrier gas He, 4 ml min^{-1}; temperature 80°C; attenuation 1 × 8. Peak identification from retention times of known compounds. 1 = Benzene; 2 = toluene; 3 = toluene-d_8; 4 = ethylbenzene; 5 = *para*-xylene; 6 = *meta*-xylene; 7 = *ortho*-xylene; 8 = *n*-propylbenzene; 9 = 1-methyl-4-ethylbenzene; 10 = 1,3,5-trimethylbenzene; 11 = 1-methyl-2-ethylbenzene; 12 = 1,2,4-trimethylbenzene; 13 = *n*-butylbenzene; 14 = 1,2,3-trimethylbenzene, From Wasik[102] with permission.

had bubbled through the stripping cell. All the aliphatic and olefinic hydrocarbon peaks are absent from the chromatogram. The aromatic hydrocarbon peaks were identified fron retention times of known compounds.

7.3.2 Flotation methods

Froth flotation

Non-saline waters. In its passage through a water column, a bubble acts as an interface between the liquid and vapour phases, and, as such, collects surface-active dissolved materials as well as colloidal micelles on its surface. Thus, in a well-aerated layer of water, the upper levels will become progressively enriched in surface-active materials. It is possible to use this effect to enrich the foamy surface layer with organic material. The foam can be as much as 200 times as concentrated in organic material as the body of the solution[103,104] and is therefore a method of preconcentrating surface-active organic materials from large volumes of water.

MacIntyre[105] suggested that adjustment of the bubble size and the depth at which the bubbles were formed might be used to control the thickness of the surface layer sampled. The collection of the charged particles ejected by bursting bubbles can be enhanced by the use of a charged glass plate as a collector.[106] A surface microlayer sampler using this principle has been devised by Fasching *et al.*[107] The literature on bubble ejection and collection has been reviewed by Blanchard.[108]

Hiraide *et al.*[109] have described a method for the rapid preconcentration of humic acid from fresh waters by coprecipitation and flotation. Flotation of suspended matter with a cationic surfactant (cetyldimethylbenzylammonium chloride) was followed, firstly, by coprecipitation with iron(III) hydroxide at pH 7 and, secondly, by flotation with anionic surfactants (sodium oleate and sodium dodecyl sulphate). This removed suspended solids and separated humic and fulvic acid. Iron(III) hydroxide was dissolved in 2 M hydrochloric acid leaving acid-insoluble humic acid to be collected by ultrafiltration. Absorption spectra, molecular weight distribution, and complexing ability were determined, after dissolving humic acid in potassium hydroxide solution. Flotation decreased the time needed for separation to a fifth of the required with normal filtration procedures.

Sea-water. When the concentrations of surface-active materials are high, the injection of bubbles into the solution from well below the surface may result in the formation of a foam at the surface. The foam can be as much as 200 times as concentrated in orgaric material as the body of the solution. Natural foams of this kind can often be seen along beaches during periods of strong winds and violent wave action. Even when the surface-active materials are not present at levels high enough for foam formation, a considerable enrichment of organic material can be found in the upper part of the water column.[110,111]

When bubbles break at the sea surface, some portion of the organic materials collected on the bubble surface is ejected into the air. The phenomenon is well known[112] but has only recently been considered as a possible method for the collection of surface-active organic materials. While most of the bubble ejection work has been done in the laboratory, a surface microlayer sampler using this principle has been devised for work in protected inlets.[113] The literature on bubble ejection and collection has been reviewed by Blanchard.[114]

7.3.3 Evaporation/distillation methods

Pervaporation

Non-saline waters. The method for separation of chlorinated hydrocarbons from dilute aqueous solutions using pervaporation membranes, has been demonstrated experimentally.[115] The mechanism of pervaporation is described as three steps: absorption of the permeating molecules at the liquid–membrane interface; diffusion of the molecule through the membranes; removal of the molecules from downstream surface of the membrane by a vacuum or carrier gas. The results indicate a viable separation can be achieved and that the solubility parameter is the most important factor for selectivity. A membrane system consisting of polyvinyl acetate as the active member on a polysulphone support has been developed and one of polyvinyl acetate sandwiched between layers of PTF has been studied.

Distillation

Non-saline waters. These have been studied by Kuo *et al.*[116] and Kozloski.[117] The latter worker condensed the distillate of a 1 litre water sample in a liquid nitrogen trap over 13 min. The purgeable components were transferred to a gas chromatograph by conventional gas purging. The recoveries and purge ratios achieved by vacuum distillation of hexane, carbon tetrachloride, chloroform, dichloromethane, benzene, 1,2-dichloroethane, diethyl ether, methyl isobutyl ketone, methyl acetate, and tetrahydrofuran were determined. Misharina *et al.*[118] have reviewed the literature between 1975 and 1985 on evaporation and other methods for preconcentrating volatile organics in natural waters.

Steam distillation

Non-saline waters. Dix and Fritz[119] examined various distillation parameters and boiling modifiers in relation to a simple steam distillation method for the extraction of organic compounds from aqueous samples. A survey mixture of 10 organic compounds, containing various functional groups, was added to water and the distillate collected for capillary gas chromatographic analysis. Recoveries of these 10 compounds collected in cold acetone, at fast-boil or with the addition of salt were tabulated. Almost all types of organic compounds with boiling points ranging from 77–238 °C showed recoveries above 90 per cent. Many azeotropes with boiling points below that of water distilled over. The addition of salt to the boiling mixture permitted the formation of an azeotrope richer in phenol.

Freeze drying

Non-saline waters. A limited amount of work has been carried out on freeze concentration methods for organics in non-saline and saline waters. The method has been applied to the preconcentration of amino acids, carbohydrates, citric acid, and phenylalanine[120] giving virtually complete recovery of these substances at concentrations up to 0.2 mg litre^{-1}. The technique has been used in conjunction with Sephadex gel chromatography to characterize organic components in the molecular weight range 200 to 5000[121] in secondary sewage effluents. The filtered samples were concentrated using a freeze drying unit with methyl alcohol as the bath fluid. The organic carbon recovery in the freeze dried samples was between 92 and 96 per cent.

Pankow[122] showed that quantitative freeze-trapping of mixtures of certain volatile chlorinated hydrocarbons can be done with a 30 m fused silica capillary column kept at between − 60 and − 100°C throughout. The effect of using different trapping temperatures on the chromatograph peak profiles was examined for 1,1-dichloroethane, dichloromethane, 1,1-dichloroethane, chloroform, carbon tetrachloride, trichloroethane, 1,2-dichloropropane, 2-chloroethyl vinyl ether, 1,1,2-trichloroethane, dibromochloromethane, tetrachloroethene, and chlorobenzene. Trapping at −80 °C appeared best, giving peak breadths of only 4–7 s at 8 °C min^{-1} and 110 cm s^{-1} carrier gas speed.

Sea-water The soluble organic phosphorus compounds arising from natural sources have been examined by a number of techniques, including concentration by freeze drying.[123-124] The most familiar of these compounds, adenosine triphosphate, is normally considered to be found in the particulate fraction, and is measured by the luciferin-luciferinase reaction. Approximately 100 per cent recovery of glucose and lindane at the 0.1 and 0.15 μg litre^{-1} level have been obtained from water by freeze drying.

Pocklington[125] has separated amino acids in seawater using freeze drying. The first step in his concentration of free acids from sea-water was the freeze drying of the sea-water samples. To reduce interferences in the later steps of the procedure, the sea salts were packed into a chromatographic column and washed with diethyl ether to remove non-polar compounds. The diethyl ether extract, particularly from surface water samples, quite often contained coloured materials as well as other organics. If a series of solvents of graduated polarity were passed through a sea salt column, a fractionation by polarity should be obtained. With the proper choice of solvents, a form of gradient elution could be devised which would result in a continuous, rather than a batch fractionation. Bohling[126,128] and Garrasi and Degens[128] used freeze drying, followed by extraction of the dried salts with organic solvents to preconcentrate amino acids from sea-water.

7.3.4 Precipitation methods

Freezing out

Non-saline waters. Slow freezing, with constant stirring, results in a concentration of organic materials in the solution remaining. The technique is most effective in water of low salinity.[129–132]

7.3.5 Dialysis and osmosis

Dialysis

Non-saline waters. Sodergren[133] studied the uptake of organic pollutants from natural water by solvent filled dialysis membranes in static and continuous-flow systems. The pattern of uptake of DDE, DDT, and a PCB mixture was similar for both systems under continuous-flow conditions. Some results obtained from field experiments in which the solvent-filled membranes were exposed to actual industrial waste-waters are tabulated. It was concluded that such membranes could be used to screen and predict the fate of pollutants in industrial waste-waters and aquatic ecosystems.

Osmosis

Non-saline waters. Reverse osmosis is a membrane process developed almost entirely in the past 20–25 years. The process is based on the fact that if two solutions of different concentrations are separated by a semipermeable membrane, water will move across the membrane in the direction of higher concentration. The driving force is provided by this difference in concentration. Since many of the dissolved substances will not pass through the membrane, the passage of water will eventually cease when the solutions on both sides reach the same concentration. This process of osmosis can be reversed by applying pressure to the more concentrated solution. The concentration of dissolved substances will then increase on that side of the membrane.

Although the process was initially developed for the purposes of desalination of water, the rejection of organics has also been extensively studied. A recent paper shows that reverse osmosis was one of the most promising techniques for removing refractory (non-volatile) organics of intermediate to higher molecular weight from water.[135] More study is needed to predict rejections of organics by reverse osmosis; it appears, however, that rejection is related to the distribution coefficient of a given compound between the aqueous solution and the hydrated membrane. Obviously, this is complicated by the concentration of inorganic ions that continuously change the solubility of the organic compounds in solution and the fact that a high salt concentration can affect the degree of hydration of the membrane.

The scheme used by Kopfler *et al.*[136] for isolating organics from water is shown in Fig. 19. Typically, 400 gallons (1514 litre) of potable water is collected into a covered 500 gallon stainless steel tank over a 2-day period. Soon after collection is begun, 140 mg of silver nitrate is added to retard bacterial growth. The collected water is adjusted to pH 5.5 with hydrochloric acid. This pH is in the optimum range for stability of cellulose acetate membranes. The water is subjected to reverse osmosis with an Osmonics Model 3319–558°C unit containing a spiral wound cellulose acetate membrane at a pressure of 190 lb in^{-2} and a 50 per cent conversion rate.

Immediately before entering the reverse osmosis unit, the water passes through a copper coil immersed in refrigerated bath maintained at 6 to 10 °C. The permeate (water forced through the membrane) is collected

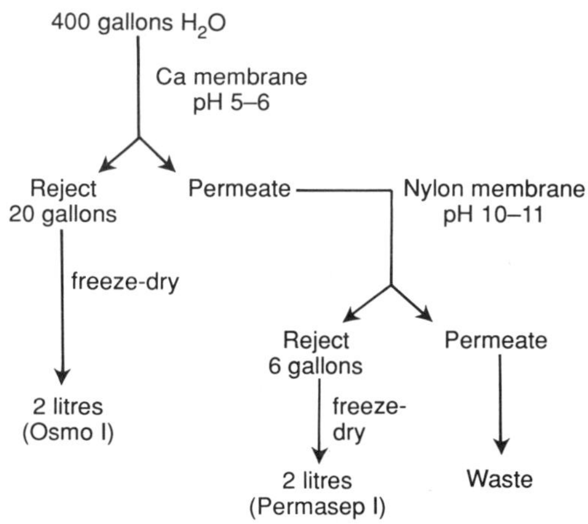

Fig. 19 Schematic for concentration of organics. From Kopfler *et al.*[138] with permission.

in a second stainless steel tank while the reject is recycled into the feed tank. When the volume of the feed has been reduced to 20 to 40 gallons, reverse osmosis is terminated. The average total organic carbon retention in this concentrate is 85 per cent. The concentrate is divided into 5 litre aliquots that are each placed in a shallow stainless steel tray. These are frozen and subsequently lyophilized. The temperature of the trays is electronically controlled and is never allowed to rise above −10 °C or reach complete dryness while in the lyophilizer. When the volume of a group of trays is sufficiently reduced, the contents are allowed to thaw and are then combined into a single tray, refrozen, and lyophilized. This procedure is continued until the total volume remaining is about 3 litre of liquid and solids.

The permeate collected from the cellulose acetate unit is again subjected to reverse osmosis in a similar manner using a Continental Water Conditioning Model 881–1 unit containing a DuPont B-9 Permasep permeator. This is an asymmetric aromatic polyamide membrane of hollow fibre configuration. It is easily degraded by the chlorine concentrations normally found in finished potable water but the chlorine is removed by the cellulose acetate membrane. This membrane can tolerate alkaline conditions and since most salts have been removed by the cellulose acetate membrane, the sample can be brought to pH 10.5 with sodium hydroxide. This will ionize phenols and acids to increase rejection by the membrane. To maintain a 50 per cent conversion rate, a pressure of 125 lb in^{-2} is maintained. The permeate is discarded and the feed solution is recycled until the volume is reduced to 2 to 3 gallons. This concentrate is frozen and lyophilized as before.

The final product obtained by lyophilizing the aqueous concentrate obtained by reverse osmosis of potable water with the cellulose acetate membrane is filtered through a coarse, sintered-glass filter to remove the precipitated salts. These salts are again lyophilized to dryness and extracted first with pentane and then with methylene chloride. The filtrate is extracted three times with pentane using 75 ml pentane per litre of solution for the first extraction and 50 ml pentene per litre for the other two extractions. The extracts are combined, dried with sodium sulphate, and concentrated in a Kuderna-Danish evaporator.[137] The solution is extracted again in the same manner with methylene chloride. It is then acidified with hydrochloric acid to pH 2 and again extracted with methylene chloride to extract acidic and phenolic compounds.

After extraction, the solvent is removed from the concentrate in a rotary evaporator at 40 °C. More organics can then be obtained by passing the concentrate through a column of XAD-2 macroreticular resin and subsequent elution with ethanol. This extraction scheme is shown in Fig. 20. The concentrate obtained with the DuPont Permasep membrane is extracted in the same manner. However, since the cellulose acetate

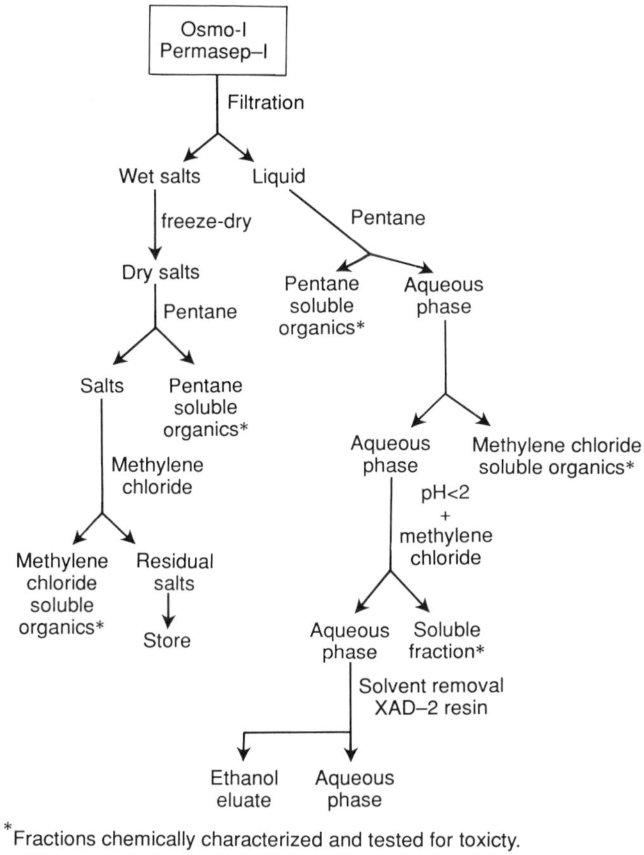

Fig. 20 Schematic for extractions of organics from reverse osmosis concentrates of tap water. From Kopfler *et al.*[138] with permission.

membrane removes most of the salts, little if any precipitation occurs on lyophilization and there is no need to extract salts separately.

Alquots of each fraction are retained for chemical analysis. The remaining portions are combined for toxicity assays. Analysis of these fractions by gas chromatography—mass spectrometry has proved the presence of some compounds not previously identified in potable water such as those shown in Table 60. Because of tar-like residues remaining in the inlet of the gas chromatographic columns, it is apparent that much of the organic residue is non-volatile and, as yet, remains unidentified.

The most volatile compounds present in the water, which are lost during the concentration procedure, were collected by stripping with helium from a water sample on to a trap containing a porous polymer. This is accomplished with the use of the system developed by Bellar and

Table 60 Organic compounds identified in reverse osmosis concentrates of Cincinnati drinking water.

Barbital	Methyl palmitate
Benzene sulphonic acid	Methyl stearate
Benzyl butyl phthalate	Methyl tetracosanoate
2-Chloroethyl methyl ether	Octadecane
Dibutyl phthalate	Pentachlorobiphenyl
Diethyl phthalate	Phenyl benzoate
Di(2-ethyl hexyl) phthalate	Phthalic anhydride
Docosane	1,1,3,3-Tetrachloroacetone
Eicosane	Tetrachlorobiphenyl
Dexadecane	Trichlorobiphenyl
Methyl docosanoate	2,4,5-Trichlorophenol

From Kopfler *et al.*[136] with permission.

Lichtenberg[138,139] modified to contain 500 ml of samples. The trap is subsequently heated, and the volatile organics back-flushed into a gas chromatograph—mass spectrometer. Typical compounds identified in potable water are shown in Table 61.

Reverse osmosis has been used for the determination in potable water of traces of organochlorine and organophosphorus insecticides,[140,142] and pentaerythritol, bis (2-chloroethyl)ether, bis(2-chloroisopropyl)ether, urea, triethylene glycol, dipropylamine, chlorobenzene, propylbenzene and benzonitrile,[141] as well as various ethers, glycols, amines, nitriles, hydrocarbons and chlorinated hydrocarbons. Cellulose acetate,[143] ethyl cellulose acetate,[142] and cross linked polyethyleneimine[141] were used as semi-permeable membranes. Chain

Table 61 Volatile organic compounds identified in Cincinnati drinking water.

Chloromethane	Dichloroacetylene
Bromomethane	Trichloethylene
Bromodichloromethane	Tetrachloroethylene
Dibromochloromethane	Bromotrichloroethylene
Chloroform	Cyanogen chloride
Bromoform	Methanol
Carbon tetrachloride	Ethanol
Chloroethane	2-Methyl propanal
Chloropropane	2-Butanone
1,2-Dichloroethane	3-Methyl butanal
Acetylene chloride	Toluene

From Kopfler *et al.*[136] with permission.

et al.[140] studied two types of reverse osmosis membranes—cellulose acetate and cross-linked polyethylenimine. With each membrane the rejection of pesticides was better than 99 per cent. Deinzer *et al.*[142] used cellulose acetate membranes.

Klein *et al.*[142] used cellulose acetate and ethylcellulose membranes and showed that the retention of a particular organic solute in potable water by a polymer membrane is related not only to the physical interactions such as size but also is dependent upon attractive forces which are normally included in the solubility of the solute in the membrane phase. These workers used solubility parameters for projections of class and specific separations with the two membranes studied.

Sea-water. Reverse osmosis has been applied to the preconcentration of organochlorine and organophosphorus insecticides[140,142] and various ethers, glycols, amines, nitriles, hydrocarbons, and chlorinated hydrocarbons in sea-water and potable water. Cellulose acetate,[143] ethyl cellulose acetate,[142] and cross-linked polyethyleneimine[141] were used as semi-permeable membranes.

7.4 Organometallic compounds

7.4.1 Electrolytic methods

Non-saline waters

Mercury. Doherty and Dorsett[144] used a copper coil preconcentrator for the electrodeposition of mercury from organomercury compounds prior to its determination by flameless atomic absorption spectrometry.

References

1. Rigin, V. I. and Rigina, I. V. (1979). *Analytical Chemistry*, *34*, 1121.
2. Vassos, B.H., Hirsch, R.F., and Letterman, H. (1973). *Analytical Chemistry*, *45* 792.
3. Torsi, G. and Palmisano, F. (1984). *Science of the Total Environment*, *37*, 35.
4. Branderberger, H. and Bader, H. (967). *Atomic Absorption Newsletter*, *6*, 101.
5. Fairless, C. and Bard, A. J. (1972). *Analytical Letters*, *5*, 433.
6. Frick, D. A. and Tallman, D. E. (1982). *Analytical Chemistry*, *54*, 1217.
7. Muhlbaier, J., Stevens, C., Graczyk, D., and Tisue, T. (1982). *Analytical Chemistry*, *54*, 496.
8. Szy, D. and Toth, A. (1970). *Mikrochimico Acta*, *5*, 960.
9. Trettenbach, J. and Henmann, K. C. (1985). *Fresenius Zeitschrift fur Analytische Chemie*, *322*, 306.

10. Wang, J. and Mahmoud, J. (1987). *Fresenius Zeitschrrift fur Analytische Chemie*, *327*, 789.
11. Prabhu, S. V., Baldwin, R. P., and Kryger, L. (1987). *Analytical Chemistry*, *59*, 1074.
12. Vandder Berg, C. M. C. (1986). *Science of the Total Environment*, *49*, 89.
13. Lundgren, G., Lundmark, L., and Johansson, G. (1974). *Analytical Chemistry*, *46*, 1028.
14. Scobbie, R. (1973). In, *Technical Topics*, Varian Techtron, Springvale, Vic., Australia.
15. Donnelly, T. H., Ferguson, J., and Eccleston, A. J. (1975) *Applied Spectroscopy*, *29*, 2.
16. Shigematsu, T., Matsui, M., Fujino, O., and Kinoshita, K. (1975) *Analytica Chimica Acta*, *76*, 329.
17. Branderberger, H. and Bader, H. (1967). *Atomic Absorption Newsletter*, *6*, 101.
18. Lund, W. and Larsen, B. V. (1974). *Analytica Chimica Acta*, *70*, 299.
19. Lund, W. and Larsen, B. V. (1974). *Analytica Chimica Acta*, *72*, 57.
20. Lund, W. and Larsen, B. V. and Gundersen, N. (1976). *Analytica Chimica Acta*, *81*, 319.
21. Lund, W., Thomassen, Y., and Doyle, P. (1977) *Analytica Chimica Acta*, *93*, 53.
22. Fairless, L. and Bard, A. J. (1972). *Analytical Letters*, *5*, 433.
23. Fairless, L. and Bard, A. .J. (1973). *Analytical Chemistry*, *45*, 2289.
24. Batley, G. E. and Matousek, J. P. (1977). *Analytical Chemistry*, *49*, 2031.
25. Jensen, F. O., Dolezal, J., and Langmyhr, F. J. (1975). *Analytica Chimica Acta*, *72*, 245.
26. Newton, M. P., Chauvin, J. V., and Davis, D. G. (1973) *Analytical Letters*, *6*, 89.
27. Newton, M. P. and Davis, D. G. (1975). *Analytical Chemistry*, *47*, 2003.
28. Dawson, J. B., Ellis, D. J., Hartley, T. F., Evans, M. E. A., and Metcalf, K. W. (1974). *Analyst (London)*, *99*, 602.
29. Torsi, G. (1977). *Analytica Chimica (Rome)*, *67*, 557.
30. Torsi, G., Oesimoni, E., Palimisano, F., and Sabbatini, L. (1981). *Analytica Chimica Acta*, *124*, 143.
31. Cowen, J. P., Hodge, V. F., and Folson, T. R. (1977). *Analytical Chemistry*, *49*, 494.
32. Reid, D. F., Key, R. M., and Schink, D. R. (1979). *Earth Planet Science Letters*, *43*, 223.
33. Breyer, P. and Gilbert, B. P. (1987). *Analytica Chimica Acta*, *201*, 33.
34. Donat, J. R. and Bruland, K. W. (1988). *Analytical Chemistry*, *60*, 240.
35. Andreae, M. O. P. I. (1981). In, *Trace Metals in Seawater, Proceedings of a Nato Advanced Research Institute on Trace Metals in Seawater*, 30/3/81–3/4/81, Sicily, Italy, eds C. S. Wong, *et al.*, Plenum Press, New York.
36. Amankwah, S. A. and Fasching, J. L. (1985). *Talanta*, *32*, 111.
37. Andreae. M. O. and Froelich, P. N. (1981). *Analytical Chemistry*, *53*, 287.
38. Foreback, C. C. (1973). In, *Some studies on the Detection and Determina-*

tion of Mercury, Arsenic and Antimony in Gas Discharges, Thesis, University of Sourth Florida, Tampa.

39. Tomkins, M. A. (1977). In, *Environmental Analytical Studies in Antimony, Germanium and Tin*, Thesis, University of South Florida, Tampa.
40. Creselius, E. A. (1978). *Analytical Chemistry*, *50*, 826.
41. Cutter, G. A. (1978). *Analytica Chimica Acta*, *98*, 59.
42. McDaniel, M., Shendrikar, A. D., Reiszner, K. D., and West, P. W. (1976). *Analytical Chemistry*, *48*, 2240.
43. Knudson, E. J. and Christian, G. D. (1973). *Analytical Letters*, *36*, 1039.
44. Orvini, E., Delfanti, R., Gallorini, M., and Speziali, M. (1981) *Analytical Proceedings (London)*, *18*, 237.
45. Shaikh, A. N. and Tallman, D. E. (1978). *Analytica Chimica Acta*, *98*, 251.
46. Hansen, E. H., Ruzicka, J., and Reitz, B. (1977). *Analytica Chimica Acta*, *89*, 241.
47. Vien, S. H. and Fry, R. C. (1988). *Analytical Chemistry*, *60*, 465.
48. Braman, R. S., Justen, J. C., and Foreback, C. C. (1972). *Analytical Chemistry*, *44*, 2195.
49. Braman, R. S. and Forebach, C. C. (1973). *Science*, *182*, 1247.
50. Bertine, K. K. and Lee, D. S. (1981). In, *Trace Metals in Seawater, Proceedings of a Nato Advanced Research Institute on Trace Metals in Seawater*, 30/3–3/4/81, Sicily, Italy, eds C. S. Wong *et al.*, Plenum Press, New York.
51. Cutter, G. A. (1978). *Analytica Chimica Acta*, *98*, 59.
52. Willie, S. N., Sturgeon, R. E., and Berman, S. S. (1986). *Analytical Chemistry*, *58*, 1140.
53. Sturgeon, R. E., Willie, S. N., and Berman, S. S. (1985). *Analytical Chemistry*, *57*, 2311.
54. Andreae, M. O. (1977). *Analytical Chemistry*, *49*, 820.
55. Braman, R. S., Johnson, D. L., Foreback, C. C. (1977). *Analytical Chemistry*, *49*, 621.
56. Howard, A. G. and Arbab-Zavar, M. H. (1981). *Analyst (London)*, *106*, 213.
57. Arbab-Zavar, M. H. and Howard, A. G. (1980) *Analyst (London)*, *105*, 744.
58. Hinners, T. A. (1980). *Analyst (London)*, *105*, 751.
59. Pierce, F. D. and Brown, H. R. (1977). *Analytical Chemistry*, *49*, 1417.
60. Talmi, Y. and Bostick, D. T. (1975) *Analytical Chemistry*, *47*, 2145. Press, New York.
61. Edmonds, J. S. and Francesconi, K. A. (1976). *Analytical Chemistry*, *48*, 2019.
62. Johnson, D. L. and Braman, R. S. (1975). *Deep Sea Research*, *22*, 503.
63. Braman, R. S., Johnson, D. L., Foreback, C. C. *et al.* (1977) *Analytical Chemistry*, *49*, 621.
64. Talmi, Y. and Bostick, D. T. (1975). *Analytical Chemistry*, *47*, 2145.
65. Braman, R. S. and Foreback, C. C. (1973). *Science*, *182*, 1247.

66. Andreae, M. O. (1977). *Analytical Chemistry*, *49*, 820.
67. Burton, J. D. (1981). In, *Trace Metals in Seawater, Proceedings of a NATO Advanced Research Institute on Trace Metals in Seawater*, 30/3–3/4/81, Sicily, Italy, ed C. S. Wong, *et al.*, Plenum Press, New York, p. 145.
68. Lee, D. S. (1982) *Analytical Chemistry*, *54*, 1182.
69. Hidalgo, J. L., Gomez, M. A., Caballero, M., Cela, R., and Perez-Bustamonte, A. (1988). *Talanta*, *35*, 301.
70. Tseng, J. H. and Zeitlin, H. (1978). *Analytica Chimica Acta*, *101*, 71.
71. Kim, Y. S. and Zeitlin, H. (1971). *Separation Science*, *6*, 505.
72. Hiraide, M., Sakurai, K., and Mizuike, A. (1984). *Analytical Chemistry*, *56*, 2851.
73. Cabezon, L. M., Caballero, M., Cela, M., and Perez-Bustamonte, J. A. (1984). *Talanta*, *31*, 597.
74. Hashimoto, T. (1971). *Analytica Chimica Acta*, *56*, 347.
75. Kim, Y. S. and Zeitlin, H. (1971). *Analytical Chemistry*, *43*, 1390.
76. Williams, W. J. and Gillam, A. H. (1979). *Analyst (London)*, *103*, 1239.
77. Smith, J. and Grimaldi, O. (1957). *Bulletin of the US Geological Survey*, *1006*, 125.
78. Voyce, D. and Zeitlin, H. (1974). *Analytica Chimica Acta*, *69*, 27.
79. Murthey, R. S. S. and Ryan, D. E. (1983). *Analytical Chemistry*, *55*, 682.
80. Kim, Y. S. and Zeitlin, H. (1972). *Separation Science*, *7*, 1.
81. Cervera, J., Cella, R., and Perez-Bustamante, J. A. (1982). *Analyst (London)*, *107*, 1425.
82. Crocker. I. H. and Merritt, W. F. (1972). *Water Research*, *6*, 285.
83. White, M. N. and Lisk, D. J. (1970). *Journal of the Association of Official Analytical Chemists*, *53*, 1055.
84. Lieser, K. H., Heuss, W. C., and Neitzen, V. (1977). *Journal of Radioanalytical Chemistry*, *37*, 717.
85. Juleli, Y. M. and Parry, S. J. (1986). *Journal of Radioanalytical and Nuclear Chemistry*, *102*, 337.
86. Gordon, C. M. and Larson, R. E. (1970). *Radiochemical and Radioanalytical Letters*, *5*, 369.
87. Jørstad, K. and Selbu, B. (1980). *Analytical Chemistry*, *52*, 672.
88. Bontron, C. and Martin, S. (1979). *Analytical Chemistry*, *51*, 140.
89. Bontron, C. (1972). *Analytica Chimica Acta*, *61*, 140.
90. Department of the Environment Water Council Standing Committee of Analysts (1982). In, *Methods for the examination of waters and associated materials. Cooper in potable waters by atomic absorption spectrometry*, RP 22 B HMC, HMSO, London.
91. Kulathilake, A. I. and Chatt, A. (1980). *Analytical Chemistry*, *52*, 828.
92. Kim, Y. S. and Zeitlin, H. (1969). *Analytica Chimica Acta*, *46*, 1.
93. Head, P. and Burton, J. D. (1970). *Journal of Marine Biological Association, UK*, *50*, 439.
94. Weiss, H. V. and Lai, M. G. (1961). *Talanta*, *8*, 72.
95. Skogerboe, R. K., Hanagan, W. H., and Taylor, H. E. (1985). *Analytical Chemistry*, *57*, 2815.

96. Hart, B. T. and Davies, S. H. R. (1977). *Australian Journal of Marine and Freshwater Research*, *28*, 105.
97. Benes, P. and Steinnes, E. (1974). *Water Research*, *8*, 947.
98. Wilson, R. L. and Di Nunzio, J. E. (1981). *Analytical Chemistry*, *53*, 692.
99. Stec, R. J., Kortyohann, S. R., and Taylor, H. E. (1986). *Analytical Chemistry*, *58*, 3240.
100. Aoyama, M., Habo, T., and Suzuki, S. (1981) *Analytica Chimica Acta*, *129*, 237.
101. Aoyama, M., Hobo, T., and Suzuki, S. (1983). *Analytica Chimica Acta*, *153*, 291.
102. Wasik, S. P. (1974). *Journal of Chromatographic Science*, *12*, 845.
103. Wallace, G. T., Jr and Wilson, D. F. (1969). In *Foam separations as a tool in chemical oceanography*, US Naval Research Laboratory Report, *1*, 6958.
104. Blanchard, F. C. (1964). *Science*, *146*, 396.
105. MacIntyre, P. (1968). *Journal Chemical Physics*, *72*, 589.
106. Blanchard, D. C. and Syzdek, L. C. (1975). *Limnology and Oceanography*, *20*, 762.
107. Fasching, J. L., Courant, R. A., Duce, R. A., and Piotrowicz, S. R. (1974). *Journal Rech. Atmosphere*, *8*, 649.
108. Blanchard, D. C. (1975). In *Applied Chemistry at protein interfaces*, Advances in Chemistry Series No. 145, ed. R. E. Baier, American Chemical Soc., Washington D. C., pp. 360–87.
109. Hiraide, M., Ren, F. L., Tamura, R., and Mizuike, A. (1987). *Mikrochimica Acta*, *416*, 137.
110. Dorman, D. C. and Lemlich, R. (1965). *Nature* (*London*), *207*, 145.
111. Karger, B. L. and DeVivo, D. G. (1968). *Separation Science*, *1*, 393.
112. Blanchard, D. C. (1964). *Science*, *146*, 596.
113. Fasching, J. L., Courant, R. A., Duce, R. A., and Piotrowicz, S. R. (1974). *Journal Rech. Atmosphere*, *8*, 649.
114. Blanchard, D. C. (1975). *Applied Chemistry at protein interfaces*, Advances in Chemistry Series, No. 145, ed. R. E. Baier, American Chemical Society, Washington DC, pp. 360–87.
115. Zhu, C. L., Yuang, C. N., Fried, J. R., and Greenberg, D. B. (1983). *Environmental Progress*, *2*, 132.
116. Kuo, P. P. T., Chian, E. S. K., and De Walle, F. B. (1977). *Water Research*, *11*, 1005.
117. Kozloski, R. P. (1985). *Journal of Chromatography*, *346*, 408.
118. Misharina, T. A., Zhuravleva, I. L., and Golovaya, R. V. (1987). *Journal of Analytical Chemistry*, *USSR*, *42*, 459.
119. Dix, K. D. and Fritz, J. S. (1987). *Journal of Chromatography*, *408*, 201.
120. Kammeren, P. A. and Lee, G. F. (1969). *Environmental Science and Technology*, *3*, 276.
121. Sachdev, D. R., Ferris, J. J., and Cleseri, N. L. (1976). *Journal of Water Pollution Control Federation*, *48*, 570.
122. Pankow, J. F. (1983). *Journal of High Resolution Chromatography and, Chromatography Communications*, *6*, 292.

123. Minear, R. A. (1972). *Environmental Science and Technology*, *6*, 431.
124. Minear, R. A. and Walanski, K. A. (1974). In *Investigation of the Chemical Identity of Soluble Organophosphorus Compounds found in Natural Waters*. Research Report UILUWRC-74-0086.
125. Pocklington, R. (1970). In *A New Method for the Determination of Amino-Acids in Sea Water and an Investigation of the Dissolved Free Amino-Acids of North Atlantic Ocean Waters*, PhD thesis, Dalhousie University, pp. 1–102.
126. Bohling, H. (1970). *Marine Biology*, *6*, 213.
127. Bohling, H. (1972). *Marine Biology*, *16*, 281.
128. Garrasi, C. and Degens, E. T. (1976). In *Analytische Methoden zur saulen-chromatographischen bestimmung von Aminosauren und Suckern im Meereswasser und Sediment. Berichte aus dem Projekt* DFGDE 74/e: 'Litoralforschung-Abwasser in Kustennahe'. DFGAbschlubkolloquim, Bremerhaven.
129. Shapiro, J. (1961). *Science*, *133*, 2063.
130. Shapiro, J. (1967). *Analytical Chemistry*, *39*, 280.
131. Habermann, H. M. (1963). *Science*, *140*, 292.
132. Kammerer, P. A., Jr and Lee, G. F. (1969). *Environmental Science and Technology*, *3*, 276.
133. Sodergren, A. (1987). *Environmental Science and Technology*, *21*, 855.
134. Morrison, G. M. P. (1987). *Environmental Technology Letters*, *8*, 393.
135. Edwards, V. H. and Schubert, P. F. (1974). *Journal of the American Water Works Association*, *66*, 610.
136. Kopfler, F. C., Melton, R. G., Mullaney, S. L., and Tardiff, R. G. (1977). *Advances in Environmental Science and Technology*, *81*, 419.
137. Gunther, F. A., Blinn, R. C., Kalbezen, H. J., Barkley, J. H., Harnis, W. D., and Simon, H. S. (1951). *Analytical Chemistry*, *23*, 1835.
138. Bellar, T. A. and Lichtenberg, J. J. (1974). *Journal of the American Water Works Association*, *66*, 739.
139. Bellar, T. A., Lichtenberg, J. J., and Kroner, R. C. (1974). *Journal of the American Water Works Association*, *66*, 703.
140. Chian, E. S. K., Bruce, W. N., and Fang. H. H. P. (1975) *Environmental Science and Technology*, *9*, 52.
141. Deinzer, H., Melton, R., and Mitchell, D. (1975). *Water Research*, *9*, 799.
142. Klein, E., Eichelberger, J., Eyer, C., and Smith, J. (1975). *Water Research*, *9*, 807.
143. Kammerer, P. A. and Lee, G. F. (1969). *Environmental Science and Technology*, *1*, 276.
144. Doherty, P. E. and Dorselt, R. S. (1971). *Analytical Abstracts*, *43*, 1887.

INDEX

acetyl cellulase membrane
complexant 39–40
acetonitrile-divinylbenzene copolymer
complexant 50
adsorbents
alumina 67
chromosorb T 48
C8 reverse phase adsorbent 72
C18 bonded silica 65–71
carbon 77–89
copper 67–8
gold 67–8
iron hydroxide 67
manganese dioxide 60–5
nucleosil 10-C18 silica 73
polypropylene film 47–8
polytetra fluoroethylene 48
porapak 71
silica 65
silica gel-celite 72–3
silica gel modified cellulose 72
silicones 48
silver 67–8, 74
thermotrap PA 72
titanium dioxide 67
tropolin C18 silica 73
zirconium dioxide 67
adsorption on solids — thermal
desorption 174–6
aliphatic hydrocarbons, solvent extraction
of 132
alkyl phthalates, solvent extraction
of 125–32
alpha scintillation counting of
lead-210 107
neptunium-237 109–10
plutonium 35–6, 102
polonium-210 107
radon-222 63
uranium-234 198
uranium-238 198
alumina adsorbent 67
amines, solvent extraction of 138, 146
aminodiacetate, cheletant 5–9, 18–32
aniline S resin, complexant 36
anodic scanning voltammetry,
determination of
cadmium 9, 199
copper 9, 199
lead 9, 199
tin 107
zinc 1, 9, 199
aromatic hydrocarbons, solvent extraction
of 132–9
atomic absorption spectrometry,
determination of
aluminium 2
antimony 66, 108, 207, 209–10
arsenic 108, 207, 209
beryllium 15–17
bismuth 66, 108, 210
cadmium 2–8, 17, 20–1, 23–5, 37–8, 77,
107, 117–18, 196, 212, 221–4
calcium 10–11
chromium 3, 10–11, 21, 24–5, 37–8,
98–101
cobalt 2–7, 10–11, 20–1, 24–5, 36–8, 66,
104, 113
copper 2–8, 10–11, 20–1, 23–5, 36–8, 77,
104, 117–18, 212, 215–18, 221–4
gallium 77
gold 67, 77, 108
indium 77
iron 3–5, 10–11, 20–1, 24–5, 37–8, 221–4
lead 2–8, 10–11, 17, 20, 23–5, 36–8,
60–2, 66–7, 102–4, 107–8, 118, 200,
210, 221–4
magnesium 10–11
manganese 3–7, 10–11, 20–1, 24–5, 37,
117–18
mercury 8, 10, 77, 107, 195–6, 215
molybdenum 8, 32, 77
nickel 2–7, 10–11, 20–1, 24–5, 36–8, 66,
77, 104, 117, 212
non-ionic detergents 138
organoantimony compounds 209
organomercury compounds 67–8, 72, 235
potassium 10–11
selenium 66, 108
sodium 10–11
tellurium 108, 210
thallium 77
tin 118, 207
uranium 8
zinc 2–8, 10–11, 17, 20–1, 23–5, 36, 118,
221–4

barium sulphate coprecipitant 106–7, 111
bentonite complexant 39
beta-spectrometry of
caesium-137 108

beta-spectrometry of calsium-137 (*cont.*)
  radon-228 106
  strontium-90 108
  yttrium 3
bis(8-quinolyl)disulphide
    coprecipitant 117–18

calcite, coprecipitant 112
calcium hydroxide coprecipitant 107–9
C8 reverse phase adsorbent 72
C18 bonded silica adsorbent 65–71
carbon adsorbent 77–89
carbon adsorption-solvent
    extraction 167–74
carboxylated polyethyleneimine–
    polymethylene isocyanate
    complexant 36
carboxylic acids, solvent extraction of 134,
    140
cathodic scanning voltammetry,
    determination of
  cobalt 199–201, 206
  copper 199–201
  lead 199–201
  molybdenum 199–201
  nickel 199–201, 206
  selenium 199–201, 206
  uranium 199–201
  vanadium 199–201
  zinc 199–201
cellulose acetate complexant 52
cellulose piperazine dithiocarboxylate
    complexant 11
cheletants
  aminodiacetate 5–9, 18–32
  chelex-100 5–9, 18–32
  chitosan 1
  immobilized 1–5
chelex-100 cheletant 5–9, 18–32
chitosan, cheletant 1
chlorinated insecticides, solvent extraction
    of 125–32, 135–6, 150
chlorinated hydrocarbons, solvent
    extraction of 135, 140, 149
chloroanisoles, solvent extraction of 138,
    146
chlorophenols, solvent extraction of 135,
    140
chromosorb T, complexant 48
chromosorb W, complexant 51
cobalt pyrrolidine dithiocarbamate
    coprecipitant 117
cocrystallization in preconcentration 220
coflotation in preconcentration of
  cadmium 212–13
  cobalt 212–13
  copper 212–13
  molybdenum 212
complexants in preconcentration
  acetyl cellulose membrane 39–40
  aniline S resin 36
  bentonite 39
  carboxylated polyethyleneimine–
    polymethylene isocyanate 36
  cellulose acetate 57
  cellulose piperazine dithiocarboxylate 11
  chromosorb W 51
  2,2-diaminodiethylcellulose 11–17, 40
  1,4-di(methacryloxymethyl)naphthalene–
    divinylbenzene copolymer 32
  dithiocarbamate substituted cellulose 32
  dithiocarbamate substituted
    chromosorb 32–3
  divinyl benzene copolymer 50
  induran loaded cellulose 18
  lipidex 5000 51
  natural cellulose 17–18
  nicotinium molybdate 10–11
  nitrocellulose membrane 39–40
  p-phenylene diaminecellulose 18
  polyacrylic acid 37–8
  polyamine polyurea resin 36
  polyethylene tri-n-octyl phosphine
    oxide 35–6
  polymethacrylic acid 50
  polystyrene-polymaleic anhydride 36–7
  poly(triaminophenyl)glyoxal 11
  polyurethane foam 33–5, 40–7, 52
  sephadex 9–10, 39, 48–9
  sep-pak C18 50
  SM-7 ester resin 37–8
  styrene–divinylbenzene copolymer 50
  sulphydryl cotton fibre 52
  thioamide substituted polyacrylonitrile
    11
  uron loaded cellulose 18
  vinyl dithiocarbamate 10
  vinyl pyridine–divinyl benzene
    copolymer 49
  Wofatit Y 77 51
copper adsorbent 67–8
coprecipitants
  barium sulphate 106–7, 111
  bis(8-quinolyl)disulphide 117–18
  calcite 112
  calcium hydroxide 107–9
  cobalt pyrrolidine dithiocarbamate 117
  diethyldithiocarbamate 113–15
  ferric hydroxide 92–102, 110–11
  8-hydroxyquinoline 113, 115, 117
  indium hydroxide 105
  iron dibenzyldithiocarbamate 116
  lead ammonium pyrrolidine 117
  lead phosphate 105–6
  lead sulphate 105–6

magnesium hydroxide 107
1,10-phenanthroline/tetraphenylboron 118
poly(vinyl pyrrolidine) 114
1-(2-pyridylazo)-2-napthol 118–19
thionalide 114, 119
thoric hydroxide 109
tris pyrrolidene thiocarbamate 113
N-vinyl-2-pyrrolidone 49–50
zirconium hydroxide 102–4, 111–12
cryogenic trapping 176–9

detection limits 77–8
dialysis, preconcentration procedures
cadmium 221–3
calcium 222
chlorinated insecticides 230
cobalt 224
copper 221–3
iron 222–3
lead 221–3
magnesium 222
nickel 224
polychlorobiphenyls 230
potassium 222
sodium 222
zinc 221–3
2,2-diamino diethylamine cellulose complexant 11–17, 40
diethlydithiocarbamate coprecipitant 113–15
differential pulse voltammetry, determination of
cadmium 98, 199
cobalt 62
copper 98, 199
lead 98, 199
molybdenum 212
selenium 199
zinc 98, 199
1,4-di(methacryloyloxymethyl)naphthalene divinylbenzene copolymer 50–1
dithiocarbamate substituted cellulose complexant 32
dithiocarbamate substituted chromosorb complexant 32–3

electrolytic preconcentration, determination of
bismuth 195
cadmium 195–7, 199–200
chromium 195
cobalt 195, 199–200, 206
copper 198–200
hydrocarbons 225–7
indium 195
iron 198
lead 195, 199–204
mercury 195–6
nickel 195, 199–200, 206
organomercury compounds 235
polonium 204–6
selenium 195, 199, 206
silver 195
tellurium 195
thallium 195
vanadium 199–200
uranium 197–200
zinc 195, 199–200
emission spectroscopy, determination of
beryllium 98
bismuth 98
cadmium 98
cobalt 98
copper 98
lead 98
nickel 98
zinc 98
evaporation preconcentration 218

ferric hydroxide, adsorbent 67
ferric hydroxide coprecipitant 92–102, 110–11
flame photometry of calcium 32
freeze drying, preconcentration of
adenosine triphosphate 229
amines 229
amino acids 229–30
antimony 218
aromatic chloro compounds 229
arsenic 218
barium 218
bromide 218
cadmium 218
calcium 218
carbohydrates 229
carboxylic acids 229
cerium 218
chlorinated hydrocarbons 229
chromium 218
cobalt 218
copper 218
ethers 229
europium 218
gold 218
haloforms 229
iron 218
lanthanum 218
mercury 218
molybdenum 218
potassium 218
scandium 218
sodium 218
strontium 220

freeze drying, preconcentration of (*cont.*)
  uranium 218, 219
  zinc 218
froth flotation preconcentration
  arsenic 215
  mercury 215
  molybdenum 215
  uranium 214
  vanadium 215
  zinc 215

gamma spectrometry of
  cadmium 7
  chromium 99–100
  cobalt 7
  iron 99 102
  manganese 7, 97
  strontium-90 220
  radon-226 62–5
  radon-228 63–5
  zinc 7
gas chromatography, determination of
  acetaldehyde 50
  acetic acid 50
  Aldrin 48
  aliphatic hydrocarbons 132, 228, 230–4
  alkyl phthalates 88, 125–32, 230–4
  amines 138, 146
  antimony 207, 209
  aromatic hydrocarbons 132–9, 228
  arsenic 207, 209
  alpha BHC 85
  benzene sulphonic acid 230–4
  carbon disulphide 147
  carboxylic acid 134, 140
  gamma Chlordane 84
  chlorinated insecticides 51, 85, 138, 146
  chloroaliphatic hydrocarbons 135, 140, 149, 228–9
  chloroanisoles 138, 146
  chlorinated aromatic hydrocarbons 229
  PP DDD 85
  PP DDE 85
  PP DDT 51
  Dieldrin 84–5
  EDTA 138, 147
  Endrin 84–5
  esters 228, 230–4
  ethers 228, 230–4
  Fluridene, herbicide 70
  Geosmin 138, 147
  germanium 207
  haloforms 135, 146, 228–9
  Heptachlor 48, 84–5
  Heptachlorexpoxide 48, 84–5
  hexafluoroacetone 50
  herbicides 136–7, 146
  hydrocarbons 125–39, 145–9, 225–8, 230–4
  ketones 225
  Lindane 48, 84–5
  linear alkyl benzene sulphonates 138, 147, 151
  methanol 50
  Methoxychlor 84
  nitrophenols 138
  nitrosamines 87, 138
  non-ionic detergents 138, 141
  organochlorine insecticides 69–70, 72–3, 83–6
  organolead compounds 152
  organomercury compounds 52, 151
  organophosphorus insecticides 136–7, 146
  organotin compounds 73, 152–3
  pentachlorophenol 51
  phenols 87, 134, 140
  phosphoric acid esters 136, 146
  polyaromatic hydrocarbons 43–7, 88, 134, 139
  polychlorobiphenyls 42, 51, 70, 86, 135, 140–6
  pyrethrins 138, 146
  selenium 209
  siloxanes 138, 147
  squoxin 138, 147
  tetrachlorodibenzo p-dioxin 51
  tetrahydrofuran 228
  tin 207, 209
  vegetable oils 138, 147
gas evolution preconcentration
  antimony 207–9
  arsenic 207–11
  germanium 207–9
  nickel 212
  selenium 209–10
  tin 207–9
glass bead immobilized preconcentration 1–5
gold adsorbent 67–8

haloforms, solvent extraction of 135, 146
headspace analysis
  of aliphatic hydrocarbons 187
  of aromatic hydrocarbons 183–4
  of chloroaromatic hydrocarbons 186–7
  of haloforms 183–4, 187
  of tetrahydrothiophen 187
  theory 179–83
haloforms, solvent extraction of 135, 146
herbicides, solvent extraction of 136–7, 146
high performance liquid chromatography of
  aliphatic diamines 71
  Atrazine 86–7
  chlorinated insecticides 72

chromate 68
haloforms 135
nitro compounds 71
organolead compounds 73
phenols 71–2
polyaromatic hydrocarbons 70–1
polychlorobiphenyls 72
Pyrazon 70
Simazine 86–7
hydrocarbons, solvent extraction of 125–33, 148–9
8-hydroxyquinoline, coprecipitant 113, 115, 117

immobilized glass bead chelation 1–5
immobilized silica chelation 1–5
indium hydroxide coprecipitant 105
Induron loaded cellulose complexant 18
inductively coupled plasma atomic emission spectroscopy, determination of
antimony 104
cadmium 24–5, 104–5, 224
chromium 21, 104–5
cobalt 21, 24–5, 104–5
copper 21, 24–5, 104–5, 224
lanthanum 104
lead 24–5, 104–5
iron 21, 24–5, 104
manganese 21, 24–5, 104–5, 224
nickel 21, 104–5, 224
phosphate 68–9
siloxanes 138
titanium 104
tungsten 77
vanadium 104
yttrium 104
zinc 21, 24–5, 104, 224
infrared spectroscopy of
hydrocarbons 40–2
vegetable oils 138
ion flotation preconcentration of
chromate 225
phosphate 225
iron dibenzyldithiocarbamate coprecipitant 116
isotope dilution spark source mass spectrometry, determination of
cadmium 24–5, 31–2
chromium 24–5
cobalt 24–5
copper 24–5, 31–2
hafnium 98
iron 24–5, 31–2
lead 24–5, 31–2
manganese 24–5
nickel 24–5, 31–2
uranium 31–2
zinc 24–5, 31–2
zirconium 98

lead ammonium pyrrolidine dithiocarbamate coprecipitant 117
lead phosphate coprecipitant 105–6
lead sulphate coprecipitant 105–6
linear alkyl benzene sulphonate, solvent extraction of 138, 147, 151
Lipdex 5000 complexant 51

magnesium hydroxide coprecipitant 107
manganese dioxide, adsorbent 60–5
8-mercaptoquinoline coprecipitant 117
molecular emission cavity analysis of selenium 94–5, 98

natural cellulose complexant 17–18
neutron activation analysis, determination of
aluminium 30, 117
antimony 30, 78, 101, 219
arsenic 30, 78, 101, 119, 209, 215, 219
barium 30, 219
bismuth 219
cadmium 26–30, 219
cerium 26–30, 78, 80–1, 219
chromium 25–6, 30, 78, 80–1, 105–6, 118
cobalt 25–6, 30, 78–81, 117–18, 219
copper 25–30, 32, 78–80, 117–18
europium 30, 78, 80–1, 219
gold 30, 78, 80–1, 219
indium 30
iron 25–6, 30, 78–81, 219
lanthanum 26–30, 78, 80–1, 219
lead 30, 32
magnesium 26–30
manganese 14–15, 25–30, 78–86, 118
mercury 30, 32, 78, 80–1, 117, 219
molybdenum 25–6, 30, 78, 117, 215, 219
nickel 25–6, 78–80
potassium 219
scandium 25–30, 78, 80–1, 219
selenium 30, 78, 219
sodium 219
strontium 30
thorium 25–6, 30
uranium 9, 25–6, 30, 78, 80–1, 117–18, 219
vanadium 25–6, 30, 117, 215
zinc 13–15, 25–30, 78–81, 117–18, 219
non-ionic detergent, solvent extraction of 138, 146
nicotinium molybdate complexant 10–11
nitrocellulose membrane complexant 39–40

nitrophenols, solvent extraction of 138
nitrosamines, solvent extraction of 138
nuclear magnetic resonance spectroscopy of
  aliphatic hydrocarbons 133
  chlorinated hydrocarbons 125
  chlorinated insecticides 136
Nucleosil 10-C18 silica adsorbent 73

octadeyl resin complexant 49–50
organomercury compounds, solvent extraction of 151
organophosphorus insecticides, solvent extraction of 136–7, 146
organotin compounds, solvent extraction of 152–153
osmosis, preconcentration procedures
  aliphatic hydrocarbons 234–5
  alkyl phthalates 234
  amines 234–5
  aromatic chloro compounds 234–5
  benzene sulphonic acid 234
  cadmium 224
  chlorinated aliphatic hydrocarbons 235
  chlorinated ethers 234
  copper 224
  esters 234
  ethers 235
  glycols 234
  manganese 224
  nickel 224
  nitriles 234–5
  organics misc 230–4
  organochlorine insecticides 234–5
  organophosphorus insecticides 234–5
  zinc 224

1,10-phenanthroline-tetraphenyl boron coprecipitant 118
phenols, solvent extraction of 134, 140
p-phenylene diamine cellulose complexant 18
phosphoric acid esters, solvent extraction of 136, 146
polyacrylic acid complexant 37–8
polyamine polyurea resin, complexant 36
polyaromatic hydrocarbons, solvent extraction of 139
polychlorobiphenyls, solvent extraction of 135, 140–6
polyethylene film, adsorbent 47–8
polyethylene-tri-*n*-octyl phosphine oxide complexant 35–6
polymethacrylicster complexant 50
polystyrene–polymaleic anhydride complexant 36–7
polypropylene film adsorbent 47–8
polytetrafluoroethylene adsorbent 48
poly(triaminophenyl)glyoxal complexant 11
polyurethane foam complexant 33–5, 40–7, 52
poly(vinyl pyrrolidine) coprecipitant 114
Poropak adsorbent 71
preconcentration of anions
  arsenate 110
  bromide 218–9
  chromate 44, 68, 110–11, 225
  dichromate 110
  phosphate 39–40, 68–9, 110, 225
  selenate 81
  selenite 81
  vanadate 113–14
preconcentration of cations
  aluminium 21, 30, 39, 98, 117
  antimony 30, 66, 78, 101, 104, 108, 207, 209, 210, 218–19
  arsenic 11–13, 30, 78, 92–4, 113–14, 207–11, 215, 218–19
  barium 30, 219
  beryllium 15–17, 98
  bismuth 35, 66, 98, 108, 195, 210
  boron 39
  cadmium 2–9, 11–15, 17, 20–1, 23–30, 37–8, 77–81, 98, 104–5, 107, 117–18, 196, 199, 212–13, 218–19, 221–4
  caesium 108–9
  calcium 26–30, 32, 218–19
  cerium 26–30, 78, 80–1, 98, 219
  chromium 3, 9–13, 21, 24–6, 30, 37–8, 78, 80–1, 98–101, 104–6, 118, 195, 218
  cobalt 2–7, 9–11, 20–6, 30, 35–8, 62, 66, 78–81, 98, 104–5, 108, 113–14, 117–18
  copper 2–13, 17, 20–32, 36–8, 77–80, 95–8, 104–5, 113–14, 117–18, 198-201, 212–13, 215–18, 221–4
  europium 30, 78, 80–1, 99, 218–19
  gallium 77, 98
  germanium 207–9
  gold 30, 67, 77, 78, 80–1, 108, 218–19
  hafnium 98
  indium 30, 67, 77, 195
  iron 3–5, 10–13, 20–1, 24–6, 30–2, 37–8, 78–81, 104, 113–14, 117, 198, 218–9, 221–4
  lanthanum 26–30, 78, 80–1, 99, 104, 218–19
  lead 3–13, 17–20, 23–5, 30–2, 36–8, 60–7, 95–98, 102–5, 107–8, 113–14, 118, 195, 199–201, 204, 210, 221–4
  magnesium 10–11, 26–30
  manganese 3–7, 9–15, 24–30, 37, 78–86, 95–7, 104–5, 113–14, 117–18, 195–7, 199–200, 224
  mercury 8–10, 13, 30, 32, 78, 80–1, 107, 113–14, 117, 195–6, 214–15, 218–19,

molybdenum 8, 25, 26, 30, 32, 77–8, 104, 117, 177, 199–201, 212, 215, 218–19
neptunium 109–10
nickel 2–7, 9–11, 20–1, 24–6, 31–2, 36–8, 66, 77–80, 95–8, 104–5, 107–8, 117–18, 145, 199–200, 206, 210, 212, 221–24
plutonium 35–6, 102
polonium 107, 109, 204–6
potassium 10–11, 218–19
radon-222 63
radon-226 62–5
radon-228 62–5, 106
selenium 1–13, 25–30, 66–108, 119, 195, 199–201, 206, 209–10, 214, 218–19
silver 36, 77–8, 80–1, 119, 195
sodium 10–11, 218–19
strontium 30, 108, 220
technetium 102, 195
thallium 195
thorium 25–6, 30
tin 33–4, 207–9
titanium 104
tungsten 11–13, 77, 119
uranium 8, 11–13, 25–6, 30–2, 78, 80–1, 109, 116–18, 198–201, 214, 219
uranium-234 197–200
uranium-238 197–200
vanadium 11–13, 25–6, 30, 95, 98, 104, 117, 199–200, 215
yttrium 3, 104
zinc 1–15, 17, 20–1, 23–32, 36, 78–81, 95–8, 104, 113–14, 117–19, 195, 199–200, 215, 219, 221–4
zirconium 95, 97, 98

preconcentration of organics
acetaldehyde 50
acetic acid 50
alcohols 166–7, 183
Aldrin 48
aliphatic diamines 71
aliphatic hydrocarbons 132–3, 228, 230–4
alkyl phthalates 88, 125–32, 230–4
amines 138, 146, 230–4
amino acids 112, 229
aromatic hydrocarbons 132–9, 183, 184, 228
Atrazine 86–7
gamma BHC 85
benzene sulphonic acid 230–4
carbamate herbicides 137
carbohydrates 229
carboxylic acids 134, 140, 229
Chlordane 84
chlorinated hydrocarbons 135, 140, 149, 228
chlorinated insecticides 51, 69–70, 72–3, 83–6, 125–32, 135–6, 146, 150
chloroanisoles 138, 146
chloroaromatics 229–34
chlorophenols 135–40
chlorophenoxyherbicides 70
Dalapon 137
p,p-DDD 85
p,p-DDE 85
p,p-DDT 51, 84, 85
Endrin 84, 85
esters 166–7, 228, 130–234
ethers 228, 230–4
EDTA 138, 147
Fluridone 70
glycols 230–4
geosmin 138, 147
haloforms 135, 146, 183–4, 229–34
Heptachlor 48
Heptachlor epoxide 48, 84–5
herbicides 70, 136–7, 146
hexafluoro acetone 50
humic acid 49, 227
hydrocarbons 40–2, 125–39, 148–9, 183–4, 225–8, 230–4
ketones 166–7, 228
linear alkyl benzene sulphonates 138, 147, 151
Lindane 48, 84–5
methanol 50
Methoxychlor 84
nitriles 230–4
nitro compounds 71
nitrophenols 138
nitrosamines 87, 138
non-ionic detergents 138, 146
organochlorine insecticides 234–5
organophosphorus insecticides 136–7, 146, 234–5
pentachlorophenols 51
phenols 71–2, 134, 140, 166–7
phenoxyacetic acid herbicides 137
phosphoric acid esters 136, 146
polyaromatic hydrocarbons 43–7, 70–1, 88, 134, 139
polychloro biphenyls 42, 51, 70, 72, 86, 135, 140–6
Pyrazon 70
pyrethrins 138, 146
siloxanes 138, 147
Simizine 86–7
Squoxin 138, 147
tetrahydrofuran 228
triazine herbicides 70
vegetable oils 138, 147

preconcentration of organometallic compounds
antimony 209
lead 73, 152
mercury 52, 67–8, 72, 151, 235
tin 73, 152–3

purge and trap analysis,
of alcohols 166–7
of aromatic hydrocarbons 169, 172–3
carbon adsorption-solvent extraction technique 162–74
of carboxylic acids 166–7, 173
of chlorohydrocarbons 166–7, 169, 172–4
of chlorinated insecticides 68
cryogenic trapping technique 176–9
of hydrocarbons 166–7, 173, 176
of esters 166–7
of ketones 166–7
of organosulphur compounds 178
of polychlorobiphenyls 166–8, 170
Tenax adsorption-thermal desorption technique 174–6
theory of 162
pyrethrins, solvent extraction of 138, 146
1-(2-pyridylazo)-2-naphthol coprecipitant 118–19

scintillation spectrometry of
bismuth-214 108
cobalt-60 108, 212
gallium 98
lead-214 108
sephadex complexant 9–10, 39, 48–9
sep-pak C18 complexant 50
SM-7 ester resin complexant 37–8
silica adsorbent 65
silica gel modified cellulose adsorbent 72
silica, immobilized chelation 1–5
silicone, adsorbent 48
siloxanes, solvent extraction of 138, 141
silver, adsorbent 67–8, 74
solvent extraction of organics
of amines 138, 146
of carboxylic acids 134, 140
of chloroanisoles 146, 158
of chlorinated hydrocarbons 135, 140, 149
of chlorophenols 135, 140
of chlorinated insecticides 125, 132, 135–6, 146, 150
continuous extractions 127
effect of pH 125
of ethylene diamine tetraacetic acid 138
evaporation losses 127
of geosmin 138, 147
of haloforms 135–46
of herbicides 136–7, 146
of hydrocarbons 125–133, 139, 148–9
of linear alkyl benzene sulphonates 138, 147, 151
of non-ionic detergents 138, 146
of nitrophenols 138
of nitrosamines 138
of organolead 152
of organomercury 151
of organophosphorus 136–7, 146
of organotin 152–3
of phenols 134, 140
of phosphoric acid esters 136, 146
of polyaromatic hydrocarbons 134, 139
of polychloro biphenyls 135, 140–6
of pyrethrins 138, 146
recovery 127
of squoxin 138, 147
from seawater 147–51
theory of 124–5
of vegetable oils 138, 147
solvent sublation preconcentration of copper 215–18
spectrofluorimetry, determination of
cerium 98
polyaromatic hydrocarbons 134
uranium 109
spectrophotometry, determination of
aluminium 39
aromatic hydrocarbons 133
arsenic 92–4
arsenate 110
bismuth 35
boron 39
carbamate insecticides 137
chromate 111, 225
chromium 3
cobalt 35
humic acids 49
organophosphorus insecticides 136
phenols 134
phenoxy acetic acid herbicides 137
phosphate 110
phosphorus 39–40
selenium 214
squoxin 138
tin 33–4
vanadium 95
zirconium 95–7
squoxin, solvent extraction of 138, 147
steam distillation preconcentration 228–9
styrene divinylbenzene copolymer, complexant 50
sulphydryl cotton fibre complexant 52

Thermotrap PA adsorbent 72
thin-layer chromatography of
chlorinated insecticides 135
chlorophenoxy herbicides 70
Dalapon 137
organochlorine insecticides 69–70
organophosphorus insecticides 136
polyaromatic hydrocarbons 43–7, 134
Triazine herbicides 70

thioamide substituted polyacrylonitrile
complexant 11
thionalide coprecipitant 114, 119
thoric hydroxide coprecipitant 109
titanium hydroxide adsorbent 67
tris-pyrrolidine thiocarbamate
coprecipitant 113
tropolin C18 silica adsorbent 73

Uron loaded cellulose complexant 18

vegetable oils, solvent extraction of 138, 147
vinyldithiocabamate complexant 10
vinylpyridine-divinylbenzene complexant 49
N-vinyl-2-pyrrolidine complexant 49-50

Wofatit Y77 complexant 51

X-ray fluorescence spectroscopy, determination of
arsenic 11-13, 113-14
cadmium 9, 11-13, 78-80
chromium 9, 11, 13
cobalt 9, 78-80, 113-14
copper 9, 11-13, 78-80, 95-7, 113-14
iron 11-13, 78-80, 113-14
lead 11-13, 95-7, 113-14, 195
lanthanum 11-13
manganese 9, 11-13, 95-7, 113-14
mercury 11-13, 113-14
nickel 9, 78-80, 95-7
organomercury compounds 52
phosphate 40
selenate 81
selenite 81
selenium 11-13, 119
tungsten 11-13
vanadate 113-14
vanadium 11-13
zinc 11-13, 78-80, 95-7, 113-14
X-ray spectrometry of
cobalt 117
copper 117
indium 67
mercury 117

zirconium hydroxide
adsorbent 67
coprecipitant 102-4, 111-12